BEYOND ROSWELL: COLD WAR COVER UPS AND CONSPIRACY THEORIES

First edition. January 10, 2026.

ISBN: 979-8233188046

Written by Christian Stefanizzi.

BEYOND ROSWELL
Cold War Cover-Ups
and Conspiracy Theories
Christian Stefanizzi

BEYOND ROSWELL
Cold War Cover-Ups
and Conspiracy Theories
Christian Stefanizzi

TOP SECRET

FOR DIRECTOR'S EYES ONLY

DECLASSIFIED
Authority: NARA
Date: ....

FROM THE DESK OF ______________________, ESQ. DATE: ___/___/20___

Headline edition July 8th NINETEEN FORTY SEVEN

The Army Air Forces has announced that a flying disc has been found and is now in possession of the Army, Army officers say the missile found sometime last week has been inspected at Roswell New Mexico and sent to Wright Field, Ohio for further Inspection.

Roswell Army Airfield (RAAF) Public information office

WALTER.G.HAUT

FROM THE DESK OF ______________________, ESQ. DATE: ___/___/20___

July 8th NINETEEN FORTY SEVEN
Initial Press Release (RAAF)

The many rumours regarding the flying disc became a reality yesterday when the intelligence office of the 509th Bomb Group of the Eighth Air Force, Roswell Army Air Field, was fortunate enough to gain possession of a disc through the cooperation of one of the local ranchers and the sheriffs office of Chaves County.

The flying object landed on a ranch sometime last week. Not knowing what it was, the rancher stored it until such time as he was able to contact the sheriffs office. The sheriff, Mr.George Wilcox, then contacted Major Jesse A.Marcel of the intelligence office.

Action was immediately taken and the disc was picked up at the Ranchers home. It was inspected at the Roswell Army Air Field and subsequently loaned by Major Marcel to higher headquarters.

Press release issued by Lt.Walter Haut, Public information Officer, Roswell Army Air Field

In his book Beyond Roswell; Cold War Cover ups and Conspiracy Theories
Christian Stefanizzi undertakes an exhaustive and methodical examinati
on of the Roswell Incident, drawing upon firsthand documentary evidence
from the FBI and the United States Army Air Forces to reassess what has
long been overlooked and what remains worthy of reinvestigation.

Mr Stefanizzi has consulted extensively with authors who have investiga
ted key aspects of the Roswell case, including Phillip Mantle, Kevin
Randle, and Nick Pope, While Roswell has been written about countless
times, the aim of this work is not to rehash familiar narratives, but
to re examine them critically, with fresh context and renewed scrutiny.

This book does not claim to solve every mystery or answer every questi
on surrounding Roswell, However, it does seek to strip away misinforma
tion, misdirection, and speculation where possible and in doing so, to
clear some of the long standing clutter that has obscured serious inquir
y into the case.

So sit back, dear readers UFOlogists, mystery enthusiasts, and seekers
of historical truth and prepare for a journey back to the origins of
the modern UFO phenomenon. Roswell has long been considered the mecca
of UFO investigation. Here, we go Beyond Roswell.

Yours Sincerely
Faithful Roswell Investigator

Acknowledgements

I would like to extend my deepest gratitude to Philip Mantle, Snow Oropesa of the Roswell UFO Museum, Kevin D. Randle, and Nick Pope for their kindness, generosity, and willingness to answer my many questions throughout this research journey. As someone still taking his first real steps into the wider field of Roswell studies and UFO research, I have been genuinely humbled by the time, insight, and expertise each of these individuals has shared with me. Their contributions have not only clarified complex areas of the Roswell case but have also encouraged me to approach the subject with a disciplined, critical, and historically grounded mind-set. These are, without question, some of the most respected and rigorously informed researchers working in ufology today. Their decades of dedication — whether through archival investigation, fieldwork, military analysis, or governmental insight — form the foundation upon which much of our modern understanding of UFO history rests. To have had the opportunity to correspond with them is something I do not take lightly. Their guidance has shaped the direction of this book, strengthened my analysis, and reminded me of the importance of balancing open-minded inquiry with intellectual honesty.

For their support, expertise, and patience, I offer my sincere thanks.

Leased Wire
Associated Press

Roswell Daily Record

RECORD PHONES
Business Office 2288
News Department
2287

Movies as Usual

Claims Army Is Stacking Courts Martial

Indiana Senator Lays Protest Before Patterson

RAAF Captures Flying Saucer On Ranch in Roswell Region

House Passes Tax Slash by Large Margin

Defeat Amendment By Demos to Remove Many From Rolls

Security Council Paves Way to Talks On Arms Reductions

No Details of Flying Disk Are Revealed

Roswell Hardware Man and Wife Report Disk Seen

Ex-King Carol Weds Mme. Lupescu

Some of Soviet Satellites May Attend Paris Meeting

Roswellians Have Differing Opinions On Flying Saucers

American League Wins All-Star Game

Miners and Operators Sign Highest Wage Pact in History

[1]

Front page of the **Roswell Daily Record**, *8 July 1947, announcing that the Roswell Army Air Field had recovered a "flying saucer" from a nearby ranch. The headline reflects the U.S. Army Air Forces' initial public statement before the story was retracted and replaced with the weather balloon explanation later the same day.*

[2]

Photographs of Alderson anthropomorphic test dummies, later cited by the U.S. Air Force as representative examples of crash-test mannequins used in high-altitude and ejection-seat experiments during the late 1940s and early 1950s.

These images have frequently appeared in official explanations offered by the Air Force—most notably in the 1994 and 1997 Roswell reports—as visual analogues for eyewitness descriptions of small humanoid bodies reportedly observed near alleged crash sites. Critics of the official explanation note that Alderson dummies were not widely deployed until several years after 1947 and were typically associated with later aerospace testing programs, raising questions about their applicability to the original Roswell-era testimony.

Illustrative reconstruction depicting a small humanoid figure escorted by military personnel, based on composite witness testimony associated with the Roswell incident. The upper image reflects a stylized rendering often circulated in UFO literature to convey reported proportions and posture of the beings described by witnesses, while the lower drawing represents a composite reconstruction derived from multiple second-hand accounts rather than a contemporaneous photograph. Such illustrations emerged in later decades as researchers attempted to visualize testimony in the absence of verified photographic evidence, highlighting the challenges inherent in reconstructing events shaped largely by memory, secrecy, and delayed disclosure.[3]

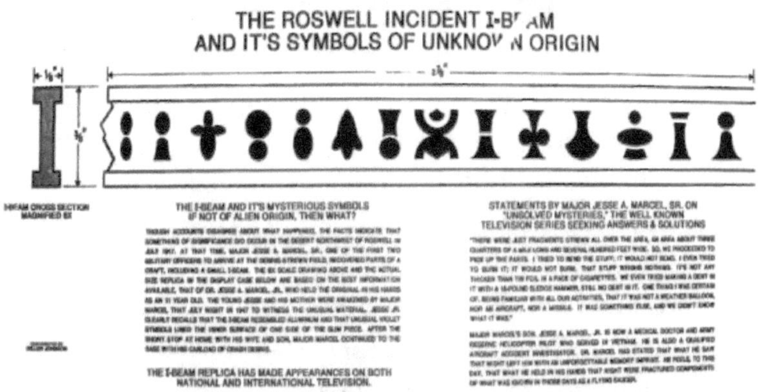

[4]

Chapter 1: The Roswell Crash (1947)

On the night of July 4, 1947, multiple witnesses across south-eastern New Mexico reported seeing a bright object streak across the sky. Among them were Floyd and Loretta Proctor, who later stated that they observed something burning as it descended, leaving a luminous trail before disappearing beyond the horizon. While early reports placed the sighting closer to the Socorro area, subsequent analysis by researchers such as Kevin D. Randle suggests that other eyewitnesses may have observed the object descending much nearer to Roswell. What is clear is that something unusual was seen in the sky that night, and its aftermath would soon become the most controversial event in UFO history. In the days that followed, debris was discovered on the Foster Ranch, northwest of Roswell, New Mexico, by rancher William W. "Mac" Brazel. Brazel described the material as scattered across a wide area of pastureland following a violent storm. Unlike the remnants of aircraft or weather balloons with which he was familiar, the debris appeared lightweight, metallic, and unlike anything he had previously encountered. Unsure of its origin, Brazel initially gathered samples and, on July 6, brought them to the Chaves County Sheriff's Office in Roswell. Sheriff George Wilcox examined the material and, recognizing its unusual nature, advised Brazel to contact the nearby Roswell Army Air Field (RAAF).

On July 7, 1947, Major Jesse A. Marcel, intelligence officer for the 509th Bomb Group—the only nuclear-capable bomber unit in the world at that time—and Captain Sheridan Cavitt of the Counter Intelligence Corps accompanied Brazel back to the ranch to inspect the debris field. Marcel later stated that the material recovered was unlike any weather balloon or conventional military hardware he had encountered in his intelligence career. According to his testimony, the debris included thin foil-like material, small beam-like structures marked with unusual symbols, and fragments that resisted burning or permanent deformation. In the early morning hours of July 8, Marcel briefly returned home before transporting the debris to Fort Worth. There, he showed portions of the material to his young son, Jesse Marcel Jr., who decades later would recall handling the debris. Marcel Jr. stated that the foil could be crumpled in the hand and would then return to its original shape, leaving no creases or damage. While this account remains anecdotal, it has been consistently maintained by Marcel Jr. throughout his life and stands as one of the most cited pieces of personal testimony associated with the Roswell incident.

Later that same day, Roswell Army Air Field issued a press release stating that personnel of the 509th Bomb Group had recovered a "flying disc," a statement that was quickly picked up by local and national media. Within hours, however, the announcement was retracted following a press briefing at Fort Worth Army Air Field, where General Roger Ramey presented debris identified as a weather balloon. Photographs of Major Marcel posed with this material became the official visual record of the incident, effectively closing the case in the eyes of the military. Yet for those directly involved—Brazel, Marcel, and others—the explanation never fully aligned with what they believed they had seen and handled. Over time, files were classified, witnesses were discouraged from speaking publicly, and the Roswell incident faded from public attention. But the contradictions remained: between eyewitness accounts and official statements, between the initial "flying disc" announcement and its rapid reversal, and between the unusual descriptions of the debris and the mundane explanation offered afterward. Roswell did not begin as a legend. It began as a reported event, documented in newspapers, military communications, and personal testimony. Whether the debris originated from an experimental military project, a classified intelligence program, or something not yet identified, the events of early July 1947 set in motion a controversy that has endured for decades. The questions raised at Roswell were not resolved—they were deferred. And it is within those unresolved spaces that the modern Roswell debate continues to exist.

Chapter 2: The Autopsy Debate (1995)

Although the Roswell incident itself occurred in 1947, public interest in the case did not remain constant. Instead, it resurfaced in waves, often triggered by new claims, documents, or media events. One of the most significant of these resurgences occurred in 1995 with the release of what became known as the "Roswell Alien Autopsy" film. Though ultimately determined to be a hoax, the film played a pivotal role in reigniting global attention on Roswell and reshaping public discourse around the case. The footage, released by British producer Ray Santilli, purported to show a post-mortem examination of a non-human body allegedly recovered from the Roswell crash. Shot in grainy black-and-white and presented as 1940s military documentation, the film was broadcast internationally and immediately divided opinion. Santilli initially claimed the footage was genuine, stating that it had been filmed by a U.S. military cameraman in 1947. Over time, however, his account shifted. He later described the film as a "recreation" based on an original that had deteriorated, and at other points suggested that fragments of authentic footage still existed. These contradictions severely undermined the credibility of the claim and ultimately contributed to the film's exposure as a fabrication.

Today, there is broad consensus among researchers, investigators, and former proponents that the autopsy film is not authentic. Special effects experts, forensic analysts, and historians have identified numerous inconsistencies, including anatomical inaccuracies, anachronistic materials, and implausible procedural details. Ray Santilli himself eventually acknowledged that the film was staged. On this point, the matter is largely settled: the Roswell autopsy film is a hoax. However, to dismiss the episode entirely would be to overlook its significance. The film was, by any reasonable standard, an effective hoax—one that arrived at a precise cultural moment. The mid-1990s marked nearly fifty years since Roswell, a period when Cold War secrecy was beginning to erode and long-classified programs were being acknowledged or declassified. Against this backdrop, the autopsy footage acted as a catalyst, reintroducing Roswell to a new generation and compelling both believers and sceptics to re-examine the historical record. In the course of my own research, including discussions with investigator and author Philip Mantle and attempted contact with Ray Santilli himself, it became clear that even those deeply involved in the autopsy controversy now view it as a deliberate fabrication rather than suppressed evidence. Mantle has consistently maintained that the film does not represent a genuine extra-terrestrial autopsy, though he, like many others, recognizes its impact in forcing renewed scrutiny of official explanations surrounding Roswell. Santilli, for his part, has declined direct engagement, leaving many questions about motivation, financing, and sourcing unanswered.

What is most revealing is not that the autopsy film was fake, but that it was so widely believed, even if briefly. Its success exposed a deep and persistent mistrust of official narratives, particularly those offered by military and intelligence institutions. That mistrust did not originate with Santilli's film; it existed long before 1995 and was merely amplified by it. The autopsy controversy demonstrated how unresolved historical gaps—missing documentation, absent physical evidence, and shifting official explanations—create fertile ground for fabrication to flourish. The Roswell autopsy film did not prove the existence of extra-terrestrial life, nor did it validate claims of recovered alien bodies. What it did prove was that the Roswell incident remains an open cultural wound. When a hoax can so effectively capture public attention, it suggests that the underlying questions have never been satisfactorily answered. In that sense, the autopsy film is best understood not as evidence, but as a symptom: a reflection of decades of ambiguity, secrecy, and contradiction surrounding Roswell. Whether intentional or opportunistic, the 1995 autopsy hoax ensured that Roswell would not remain confined to history books or archival debates. Instead, it propelled the incident back into the global consciousness, where it continues to occupy the uneasy space between fact, fiction, and unresolved history. It is within that space that the Roswell mystery persists—not because every claim is true, but because too many questions remain unanswered.

Expert Correspondence: Philip Mantle on the Roswell Alien Autopsy Film

As part of my broader investigation into Roswell-related claims, I reached out to long-time UFO researcher and film analyst Philip Mantle, whose work on the Alien Autopsy case is widely considered the most extensive to date. On 20 November 2025, I received the following responses to my questions regarding the origin, authenticity, and cultural impact of the 1995 film.

What follows is a formal, lightly edited presentation of Mantle's statements for clarity, preserving his meaning exactly as provided.

1. Authenticity of the Alien Autopsy Film

Mantle's conclusion:

> "All of it is 100% fake."

According to Mantle, no element of the film — props, footage, backstory, nor claimed 'original reels' — has any evidential basis. His position, informed by direct access to individuals responsible for creating the film, is categorical.

2. Ray Santilli's Changing Claims

Mantle attributes Santilli's shifting explanations to the pressure of scrutiny:

> "Santilli changed the story to suit what was happening. Every time something was uncovered and cast doubt on his claims he would change things yet again."

This shifting narrative, Mantle argues, reflects adaptation rather than authenticity.

3. Spyros Melaris and the Creation of the Film

Mantle identifies Spyros Melaris as the creative force behind the film:

> "Spyros is the designer, editor and filmmaker who came up with the idea, designed it and filmed it. It is really that simple. He is also the copyright holder."

Mantle also notes he personally reviewed Melaris's supporting documentation:

> "I have been to Spyros's house and not just heard him tell me his story but back it up with documentary evidence."

The implication is that the hoax's construction is thoroughly documented — leaving no room for a genuine original.

On Santilli's claim of 'damaged real footage,' Mantle writes:

> "He claims all kinds of things but cannot support such claims."

4. Purpose and Motivation Behind the Hoax

Mantle distinguishes between the intentions of Melaris and Santilli:

Melaris:

> "The film was created by Spyros as an example of how people can be fooled. As a magician he wanted to show the world the greatest magic trick and then reveal it."

Santilli:

> "He lives to make money... from the very outset this was a commercial project."

Mantle's view is that the hoax began as a demonstration of illusion but was commercialised aggressively by Santilli.

5. Impact on Roswell Research

Mantle acknowledges the film amplified public interest:

> "When the film was released, it caused a great deal more interest in the Roswell crash. Books were sold, conferences were sold out and even the X-Files had their own alien autopsy episode."

However, he also implies that its sensational nature complicated serious investigation, blurring the line between folklore, entertainment, and historical inquiry.

Closing Reflections on the Alien Autopsy Film

After corresponding directly with Philip Mantle, studying his research, reading Beyond Roswell, reviewing additional sources, and examining the twenty-minute film produced by Ray Santilli and Spyros Melaris, I believe I can reach a clear conclusion. Mantle—who has investigated the Alien Autopsy case more deeply than anyone—demonstrates convincingly that the film is a fabrication. His answers to my questions were unambiguous: the footage, the props, the set, and the entire narrative surrounding the so-called "cameraman Jack" were manufactured. Crucially, Santilli has never produced a single frame of genuine original footage, despite decades of opportunity to do so. His constant revision of the story further weakens the credibility of his claims. Given all the evidence, and Philip's first-hand documentation from Spyros Melaris—the man who designed, directed, and filmed the entire sequence—I conclude that the Alien Autopsy film is indeed a hoax. It appears to have served two purposes: for Santilli, a commercial enterprise; for Melaris, a kind of elaborate magic trick intended to demonstrate how easily the public can be fooled. I attempted to contact Santilli myself and received no response. Without verifiable original film, without a real cameraman, and with the creators openly acknowledging their roles, there is no remaining justification to treat the footage as authentic. This does not, however, erase the broader mystery of Roswell. If anything, the hoax distorts and distracts from the genuine historical questions raised by the 1947 incident. But in terms of the Alien Autopsy film itself, after careful investigation, I consider the matter closed: the film is not evidence of extra-terrestrial life, but evidence of how powerful misinformation and spectacle can be.

Chapter 3: Roswell in the Atomic Age

Any serious examination of the Roswell incident must begin with the historical moment in which it occurred. July 1947 was not simply the early post-war period; it was a transitional threshold in human history. Less than two years had passed since the detonation of atomic weapons over Hiroshima and Nagasaki. For the first time, humanity possessed the means to destroy entire cities in a single act. This shift fundamentally altered military planning, intelligence priorities, and the culture of secrecy that governed them. The Roswell Army Air Field was not an ordinary installation. It was home to the 509th Bomb Group—the only military unit in the world equipped to deliver operational atomic weapons. In 1947, this made Roswell one of the most strategically sensitive locations on the planet. Security protocols at such installations were not merely defensive; they were anticipatory. Information control was embedded in daily operations, and deviations from routine were treated with heightened seriousness.

This context matters because it establishes why Roswell could not have been treated as an open or transparent incident regardless of the nature of what was recovered. Any unexplained object descending near the 509th—whether experimental, foreign, or unknown—would have triggered immediate classification, restricted access, and centralized command authority. Secrecy was not an anomaly; it was the default condition. It is also within this period that reports of unidentified aerial phenomena began to increase sharply. Sightings clustered not randomly, but disproportionately near nuclear facilities, weapons laboratories, and strategic air bases. This pattern would persist throughout the Cold War, recurring at missile sites, bomber bases, and test ranges across the United States and, later, the Soviet Union. Whether interpreted as coincidence, misidentification, surveillance, or something more anomalous, the correlation itself is historically documented and cannot be dismissed.

The Cold War environment further reinforced institutional secrecy. The United States entered a prolonged period in which information was compartmentalized, records were routinely classified, and public disclosure was subordinated to perceived national security. In this climate, explanations were often designed not to satisfy curiosity but to stabilize public confidence and minimize uncertainty. This does not imply fabrication by default, but it does explain why official narratives were frequently simplified, incomplete, or revised. Importantly, this context explains how Roswell was managed—but not what occurred. Strategic sensitivity accounts for the rapid retraction of the initial press release, the consolidation of authority under senior command, and the disappearance of primary records. It does not, by itself, resolve the discrepancies in witness testimony, material descriptions, or timeline inconsistencies that followed. Understanding Roswell as an event embedded within the atomic and early Cold War era allows for a more disciplined analysis. It removes the false choice between naïve transparency and conspiratorial intent. The secrecy surrounding Roswell was not extraordinary; it was structurally inevitable. The unresolved question is not why information was controlled, but whether the explanations that replaced it were sufficient. That question remains open—and it is the space in which the Roswell case continues to reside.

The 509th Bomb Group: Strategic Context and Unavoidable Attention

One of the least disputed facts in the Roswell record is also one of the most significant: in 1947, Roswell Army Air Field was home to the 509th Bombardment Group, the only military unit in the world equipped with operational atomic weapons. This was not merely another bomber group. The 509th represented the cutting edge of post-war military power and the most closely guarded capability on the planet. Just two years earlier, aircraft from this unit had delivered atomic bombs to Hiroshima and Nagasaki, permanently altering the nature of warfare. In the early Cold War environment, the presence of the 509th made Roswell an extraordinary location. Security was paramount, secrecy routine, and sensitivity to any unexplained aerial activity understandably heightened. Any incident occurring near the 509th—regardless of origin—would have been treated as a potential intelligence threat until proven otherwise.

This context matters when assessing why the Roswell incident unfolded the way it did. A crash or debris recovery near the world's only atomic strike force would not have been handled as a curiosity. It would have triggered immediate containment, rapid classification, and strict control of information. From a military standpoint, even the appearance of foreign surveillance near the 509th would have been unacceptable, regardless of whether that surveillance was real, misunderstood, or entirely mundane. In the decades following Roswell, researchers such as Richard Dolan have documented a recurring association between unidentified aerial phenomena and nuclear infrastructure, including weapons laboratories, missile fields, and test ranges. While such correlations do not establish causation, they suggest that nuclear facilities have repeatedly been focal points for anomalous reports. This pattern has been observed not only in the United States, but also in the Soviet Union and other nuclear-capable states, reinforcing the idea that strategic assets tend to attract heightened attention—human or otherwise.

It is not necessary to assume extra-terrestrial intent to recognize this dynamic. Nuclear installations are among the most heavily monitored sites on Earth, drawing constant aerial, electronic, and intelligence scrutiny. During the late 1940s, when radar systems were new and identification protocols incomplete, unusual readings or unfamiliar objects could easily generate alarm, misinterpretation, or exaggerated response. From this perspective, Roswell does not require an extraordinary motive to explain the extraordinary secrecy that followed. The combination of an unexplained event, proximity to the 509th, and the political climate of 1947 was sufficient to justify an immediate and forceful containment strategy. Once an initial press release escaped control, the rapid substitution of a more mundane explanation—whether balloon debris or otherwise—served the practical purpose of restoring security and public calm.

Seen this way, Roswell is less a story about why something came to New Mexico, and more a story about why nothing was allowed to remain unexplained near the most sensitive military unit on Earth. The presence of the 509th did not make Roswell unique because it invited attention, but because it guaranteed that any anomaly—real or perceived—would be buried under layers of classification. Whether unidentified phenomena are ultimately explained as misidentifications, advanced technology, or something still unknown, the strategic reality remains the same: Roswell was the last place in 1947 where ambiguity could be tolerated. In that sense, the 509th Bomb Group did not cause the Roswell mystery—but it ensured that whatever happened there would never be handled transparently.

Chapter 4: The "Missing Nurse" Revisited — A Process of Elimination

Claims surrounding a so-called "missing nurse" at Roswell have circulated for decades, often blending verifiable personnel records with speculative narrative. To assess these claims responsibly, it is necessary to distinguish documented individuals from unverified accounts and to identify where overlap exists—and where it demonstrably does not. One frequently cited candidate is 1st Lt. Eileen M. Fanton, who was assigned to the Roswell Army Air Field (RAAF) Station Hospital from 26 December 1946 to 4 September 1947. Fanton's service record is well documented. She was promoted to First Lieutenant in June 1947, served as a general duty nurse, and later completed an overseas assignment in England. Of the five nurses assigned to the Roswell station hospital during July 1947, Fanton was the only one whose subsequent service in England can be independently confirmed—an unusual detail that has drawn sustained attention from researchers.

Several of Fanton's biographical characteristics align superficially with elements found in the "missing nurse" narrative: she was single in 1947, physically petite (approximately 5'1" and 100 pounds), dark-haired, dark-eyed, and Roman Catholic. Her confirmed later service in England, including assignment to the 7510th USAF Hospital (APO 240), further explains why she became a focal point in attempts to identify the nurse described in Alien Interview.

However, this comparison has clear limits.

Eileen Fanton died in 1975. By contrast, the individual known as Matilda O'Donnell MacElroy, as described in Lawrence R. Spencer's Alien Interview, is claimed to have been alive into the late 1990s, with correspondence allegedly exchanged as late as 1998. On chronology alone, Fanton cannot be MacElroy.

This is not a minor discrepancy. It is decisive.

Fanton's elimination matters because she has frequently been presented—explicitly or implicitly—as the solution to the "missing nurse" question. In reality, she functions as a control case: a fully documented Roswell nurse whose life trajectory demonstrates how easily biographical overlap can be mistaken for identity. Removing Fanton from consideration strengthens the analysis rather than weakening it. What remains unresolved are the other four nurses assigned to the Roswell Army Air Field Station Hospital during July 1947. Unlike Fanton, these individuals have received little sustained scrutiny. Their post-Roswell assignments, overseas service (if any), discharge dates, and dates of death have not been systematically cross-referenced against claims made in Spencer's narrative. As a result, the record remains incomplete rather than closed. It is equally important to clarify what this comparison does—and does not—suggest. The elimination of Fanton does not validate Spencer's account, nor does it confirm the existence of Matilda O'Donnell MacElroy. Spencer himself has stated that Alien Interview should be treated as fiction, and no independent military, medical, or archival records corroborate MacElroy's claims. At the same time, the inability to definitively match MacElroy to a known Roswell nurse highlights a persistent gap in the historical record rather than resolving it.

In this sense, the "missing nurse" problem mirrors the larger Roswell issue. It is not that extraordinary claims are proven by the archive; it is that the archive remains incomplete. Some records exist, others are absent, and some may never have been created or released. The temptation to close the case by assigning a convenient identity should be resisted in favour of acknowledging what can—and cannot—be shown.

Closing Assessment: The Nurse Question and the Limits of Identification

The comparison between Eileen M. Fanton and the figure known as Matilda O'Donnell MacElroy establishes one essential point: they are not the same individual. Fanton's documented service record and her death in 1975 conclusively exclude her as a candidate for the nurse described in Alien Interview.

What remains unresolved is whether any of the five nurses assigned to Roswell in July 1947 could plausibly align with Spencer's narrative. Of those five, only Fanton has been identified with sufficient certainty to permit firm conclusions. The remaining four nurses have not been comparably traced in the public record.

This absence matters. Spencer's narrative appears to draw selectively from authentic service profiles—female military nurses, post-war silence, overseas duty—while extending beyond verifiable history. Whether this reflects deliberate fiction, symbolic storytelling, or narrative reconstruction cannot be determined from available evidence. What can be determined is that retroactively assigning the MacElroy identity to Fanton oversimplifies a more complex evidentiary landscape.

The responsible conclusion is therefore neither affirmation nor dismissal. The "missing nurse" narrative belongs in a narrow analytical category: an unresolved artefact arising where documentation thins and speculation rushes in to fill the void. Historical integrity requires acknowledging that distinction.

Appendix A

Nursing Staff Assigned to Roswell Army Air Field Station Hospital — July 1947

(lightly tightened for consistency; content unchanged)

1. 1st Lt. Eileen M. Fanton

Status: Documented / Eliminated as MacElroy candidate

Known:

- Assigned to RAAF Station Hospital: 26 Dec 1946 – 4 Sept 1947
- Rank: First Lieutenant (promoted June 1947)
- Role: General duty nurse
- Overseas service: England (confirmed)
- Date of death: 1975

Conclusion: Cannot be Matilda O'Donnell MacElroy.

2–5. Remaining Nurses (Names unconfirmed in public record)

Status: Inconclusive

Known: Assigned to RAAF Station Hospital during July 1947

Unknown: Later assignments, overseas service, discharge dates, dates of death

Conclusion: Insufficient data for confirmation or exclusion.

References — Chapter 4

Primary & Government Sources

United States Air Force. The Roswell Report: Case Closed. Washington, D.C., 1997.

United States Air Force personnel summaries referencing nursing staff at Roswell Army Air Field, 1947 (secondary citations via GAO and USAF reports).

Books & Scholarly Works

Randle, Kevin D. The Roswell Encyclopaedia. Avon Books, 1999.

Randle, Kevin D. The Roswell UFO Crash Update: 1995. HarperCollins, 1995.

Friedman, Stanton T., and Don Berliner. Crash at Corona. Marlowe & Company, 1992.

Launius, Roger D. "Roswell and the Cultural Afterlife of UFOs." Smithsonian National Air and Space Museum essays.

Contested / Literary Sources

Spencer, Lawrence R. Alien Interview. New Era Publications, 2008.

(Author states the work should be treated as fiction; included here solely for comparative analysis.)

Archival & Secondary Research

International UFO Museum and Research Center, Roswell, NM — nurse-related testimony files.

GAO Report: Results of a Search for Records Concerning the 1947 Crash Near Roswell, New Mexico (1995).

[5]

[6]

Chapter 5: The Dummy Problem: Contradictions in the Official Record

When the U.S. Air Force released The Roswell Report—Case Closed in 1997, it was presented as the definitive resolution to decades of claims regarding recovered bodies at Roswell. According to the report, witnesses who believed they had encountered non-human entities were actually recalling encounters with anthropomorphic test dummies used in high-altitude parachute experiments during the 1950s. The Air Force argued that memory distortion—specifically the retroactive misplacement of later events into the summer of 1947—accounted for the persistence of body-related testimony. At first glance, the explanation appears reasonable. Human memory is imperfect, and Cold War-era test programs often involved unfamiliar equipment and secrecy. However, when the dummy explanation is examined closely, it introduces a series of contradictions that undermine its effectiveness as a comprehensive resolution. Rather than closing the case, it highlights significant inconsistencies in chronology, geography, anatomy, and witness testimony.

A Timeline That Does Not Hold

The most immediate problem with the dummy explanation is chronological. Anthropomorphic test dummies were not introduced into U.S. military testing until 1953, with projects such as High Dive and Excelsior conducted primarily between 1953 and 1960. These tests were centred at Holloman Air Force Base in New Mexico—not at Roswell Army Air Field—and occurred six or more years after the events of July 1947. This creates an insurmountable temporal gap. Dummies cannot explain reports made in 1947, nor can they account for testimony recorded before the dummy programs existed or were publicly known. To bridge this gap, the Air Force explanation requires a broad and uniform displacement of memory across multiple witnesses, spanning several years and applied consistently to individuals from different backgrounds and contexts. Such a mechanism is theoretically possible but historically implausible when applied at this scale.

The Witness Problem

The dummy explanation relies heavily on the assumption that witnesses misremembered later experiences as earlier ones. Yet many Roswell witnesses anchored their accounts to specific dates, locations, and circumstances tied directly to the summer of 1947. These include Major Jesse A. Marcel, his son Dr. Jesse Marcel Jr., rancher William Brazel, and mortician Glenn Dennis, among others. Dr. Marcel Jr., who was a child in 1947, consistently described unusual materials and body-related elements explicitly linked to his father's role in the original incident. His recollections do not resemble 1950s parachute recovery operations and were not shaped by later military test environments. Glenn Dennis's testimony—controversial though it remains—also places alleged body-related events within the Roswell Army Air Field hospital context of 1947, not at Holloman Air Force Base years later. The dummy explanation does not adequately address these contextual specifics. It assumes a generalised confusion while ignoring the precise situational details provided by multiple witnesses.

The Fulgham Explanation

In an effort to account for reports of "alien" appearances in medical settings, the Air Force later advanced the case of Captain Dan D. Fulgham. On May 21, 1959, Fulgham sustained a serious head injury after striking a balloon gondola, resulting in a pronounced hematoma—a localised swelling that caused severe facial distortion. According to the testimony of Joseph Kittinger, Fulgham's face swelled rapidly, closing both eyes and rendering his appearance, in Kittinger's words, "grotesque." There is no dispute that Fulgham was badly injured, nor that such trauma could shock observers. What remains unclear is how this isolated incident can reasonably be extended to account for earlier or independent testimony describing non-human bodies recovered in connection with Roswell. Fulgham was a fully grown adult male. Even accounting for extreme swelling, his stature, limb proportions, and overall body size bear no resemblance to descriptions consistently placing the bodies at between 3.5 and 4.5 feet tall. More importantly, the Fulgham explanation addresses only facial distortion. It does not account for recurring descriptions of non-human anatomical features such as disproportionate head-to-body ratios, elongated arms, unusual hands, or non-standard skin texture. Nor does it explain reports involving multiple bodies, guarded transport, or sealed containment. As such, the Fulgham incident may explain how a severely injured airman could alarm observers in a medical setting, but it does not resolve the broader body of testimony associated with Roswell.

Anatomical Mismatch

Even setting chronology aside, the physical descriptions reported by witnesses do not align with anthropomorphic test dummies. The dummies used in the 1950s were designed to approximate adult human proportions. They typically stood five to six feet tall and featured full-sized torsos, limbs, and heads consistent with human biomechanics.

By contrast, witnesses describing bodies associated with Roswell repeatedly reported:

Heights between approximately 3½ and 4½ feet

Disproportionately large heads

Thin limbs and narrow torsos

Non-standard hand structures

These features do not correspond to test dummies, either in design intent or execution. The suggestion that such pronounced differences could be overlooked or conflated strains credibility.

Why the Dummy Explanation Emerged

Notably, the dummy narrative did not appear until the mid-1990s—nearly fifty years after Roswell and only after body-related testimony had become widespread. Earlier official explanations focused on weather balloons and Project Mogul. The introduction of test dummies appears less like the discovery of new evidence and more like a retrofitted solution to a growing evidentiary problem. Researchers such as Richard Dolan and Kevin D. Randle have pointed out that this layered approach—balloons first, then classified projects, then dummies—undermines the coherence of the official narrative. Rather than converging on a single explanation, the story expands, with each new element addressing a different set of claims without reconciling them into a consistent timeline.

Memory as a Catch-All Explanation

The Air Force ultimately relies on memory distortion to bridge these gaps. While memory error is real and well documented, it cannot reasonably account for:

Multiple independent witnesses making similar claims

Consistent anatomical descriptions

Accurate placement of events within the 1947 timeframe

The absence of confusion between Roswell and Holloman facilities

Invoking memory failure selectively—only where testimony conflicts with official explanations—turns psychology into a catch-all dismissal rather than an analytical tool.

Kevin Randle's Challenge

Veteran Roswell researcher Kevin D. Randle has gone further than merely questioning the Air Force's conclusions. Based on decades of archival work, he has argued that the balloon and dummy narratives functioned less as clarifications than as deliberate misdirection's. Randle has noted that radar-reflector targets cited in the Project Mogul explanation were used primarily on the East Coast and may not have been deployed in New Mexico launches relevant to Roswell.

Randle's position does not require acceptance of extra-terrestrial conclusions. His argument is narrower and more troubling: that the official explanation does not accurately reflect the historical record. If true, this suggests Roswell was managed as a public relations problem rather than resolved as an evidentiary one.

Conclusion: An Explanation That Explains Too Little

By attempting to retrofit 1950s technology into a 1947 event, the Air Force weakened its own case. Anthropomorphic test dummies cannot explain reports from a time when those dummies did not exist, in places where they were not used, involving descriptions they do not match. This does not prove that alien bodies were recovered at Roswell. But it does demonstrate that the dummy explanation fails as a comprehensive resolution. Rather than closing the case, it highlights a deeper problem: the official record has repeatedly adapted to testimony instead of anticipating it. Roswell remains unresolved not because witnesses refuse to forget, but because explanations continue to arrive out of sequence

References & Further Reading — Chapter 5

The Dummy Problem: Contradictions in the Official Record

Primary Government Sources

United States Air Force. The Roswell Report—Case Closed. Washington, D.C.: Department of the Air Force, 1997.

The official Air Force explanation attributing reported bodies to anthropomorphic test dummies used in high-altitude experiments during the 1950s.

United States Air Force / Holloman Air Force Base. A Historical Look at Anthropomorphic Dummies. Holloman AFB archives, 1953–1960.

Documents the development, purpose, and deployment timeline of test dummies under Projects High Dive and Excelsior.

United States General Accounting Office (GAO). Results of a Search for Records Concerning the 1947 Crash Near Roswell, New Mexico. Washington, D.C., 1995.

Confirms the absence of documentation placing anthropomorphic test dummies in use prior to 1953.

Eyewitness & Affidavit Sources

Marcel, Jesse A., Jr., and Linda McGuire. The Roswell Legacy. New York: Berkley Books, 2007.

First-hand childhood account of debris and materials shown by Major Jesse Marcel Sr. in July 1947.

Marcel, Jesse A., Sr. Interviews conducted by Stanton T. Friedman, 1978–1980.

Statements describing unusual debris, symbols, and material properties inconsistent with conventional aircraft or balloon components.

Haut, Walter G. Affidavit, December 26, 2002.

Former Public Information Officer at Roswell Army Air Field, stating his belief that non-human bodies were recovered.

Dennis, Glenn. Interviews archived at the International UFO Museum and Research Center, Roswell, New Mexico.

Accounts of conversations with base personnel regarding bodies, coffins, and hospital activity in July 1947.

Rowe, Frankie D. Testimony cited in Kevin D. Randle, The Roswell UFO Crash Update: 1995.

Childhood recollections passed down from her father, a Roswell firefighter, concerning debris and "small bodies."

Technical & Medical Context

Kittinger, Joseph W. Testimony regarding the 1959 balloon gondola accident involving Capt. Dan D. Fulgham.

Descriptions of Fulgham's facial injuries and swelling following impact during high-altitude balloon testing.

Fulgham, Dan D. Medical incident records referenced in The Roswell Report—Case Closed (1997).

Used by the Air Force to suggest hospital misidentification of injured personnel as "alien bodies."

Scholarly & Investigative Analyses

Randle, Kevin D. The Roswell Encyclopaedia. New York: Avon Books, 1999.

Comprehensive reference work detailing witnesses, timelines, and contradictions in official explanations.

Randle, Kevin D. The Roswell UFO Crash Update: 1995. New York: HarperCollins, 1995.

Addresses the emergence of the dummy explanation and evaluates its consistency with earlier testimony.

Dolan, Richard M. UFOs and the National Security State, Vol. I. Charlottesville, VA: Hampton Roads Publishing, 2000.

Places the dummy narrative within a broader pattern of evolving official explanations and Cold War secrecy.

Pflock, Karl T. Roswell: Inconvenient Facts and the Will to Believe. Amherst, NY: Prometheus Books, 2001.

A sceptical analysis that nonetheless acknowledges the chronological and anatomical problems with the dummy explanation.

Psychological & Memory Studies (Contextual)

Schacter, Daniel L. The Seven Sins of Memory. New York: Houghton Mifflin, 2001.

Provides a framework for understanding memory distortion, frequently cited by proponents of the dummy explanation.

Neisser, Ulric. "Memory: What Are the Important Questions?" American Psychologist 47, no. 1 (1992).

Useful for evaluating claims of retroactive memory displacement across long time spans.

Author's Note on Sources

This chapter distinguishes between documented historical timelines and retrospective psychological explanations. While memory distortion is a valid scientific concept, the references cited here demonstrate that the anthropomorphic dummy explanation fails to account for the chronology, physical descriptions, and contextual specificity of Roswell-era testimony. The purpose of this reference list is not to assert extra-terrestrial conclusions, but to show that the official explanation remains internally inconsistent when examined against its own sources.

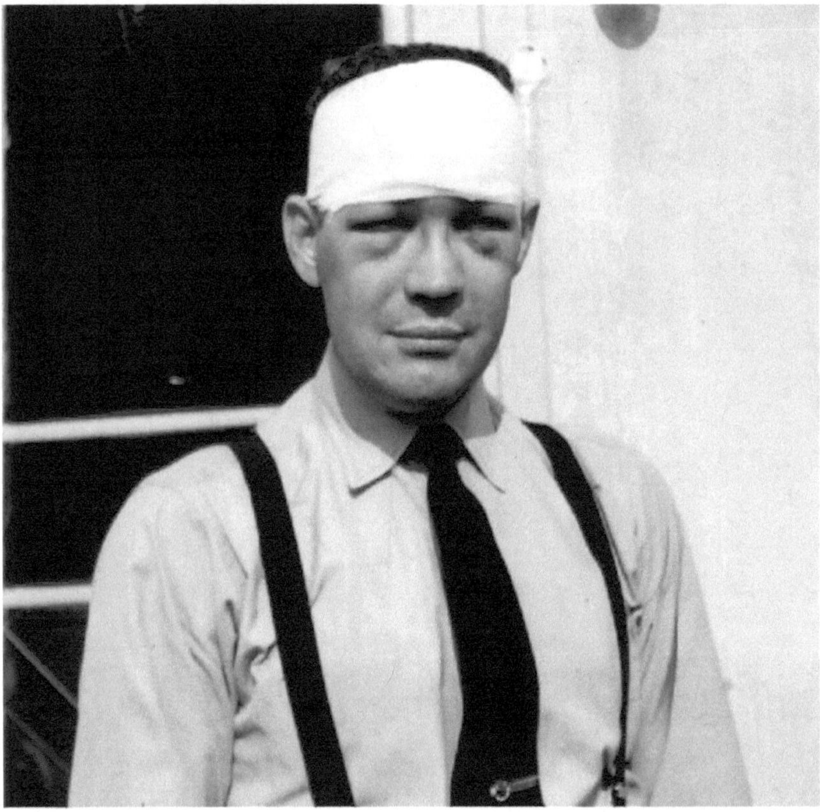

[7]

Major Dan.D.Fulgham The Air Force's explanation for the misidentification of the Alien head, due to the hematoma which caused his face and eyes to swell and deform.

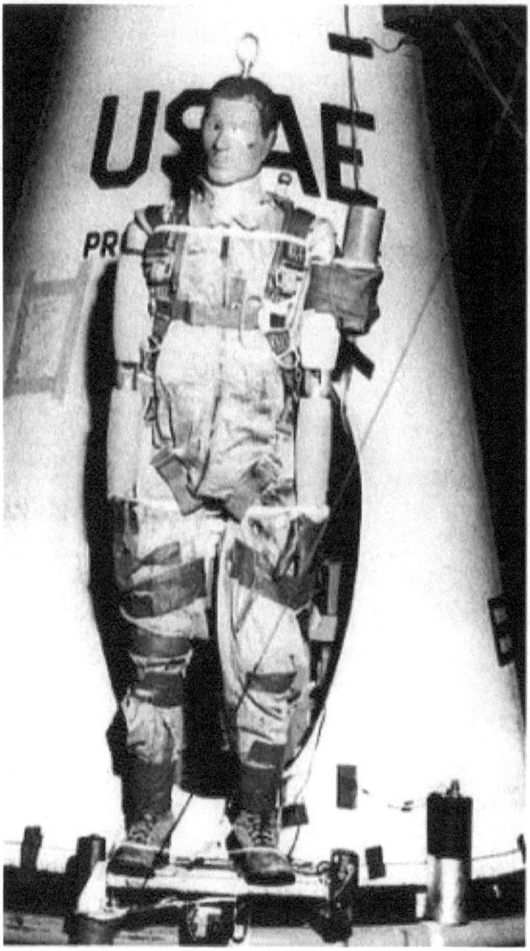

[8]

Anthropomorphic test dummies used in U.S. Air Force high-altitude research programs (1950s).

Left: A fully instrumented dummy prepared for drop testing.

Right: Recovery personnel with dummies following a test.

These images were reproduced in The Roswell Report: Case Closed (U.S. Air Force, 1997) as part of the official explanation attributing Roswell-era "body" reports to later test-dummy operations conducted years after 1947.

Chapter 6: Managing Silence — Classification, Confusion, and Control

By the summer of 1947, the United States military had emerged from the Second World War as the most powerful and technologically advanced force on Earth. Yet it had also inherited a system of secrecy shaped by wartime urgency and reinforced by the early Cold War. Understanding Roswell requires understanding this system—not as a conspiracy, but as an administrative culture in which silence, compartmentalization, and ambiguity were standard operating procedure. In 1947, there was no Freedom of Information Act, no expectation of public transparency, and little tolerance for press scrutiny of military affairs. The default posture of the armed forces was not disclosure, but non-engagement. Information flowed upward through narrow channels, while lateral communication was restricted and public explanation discouraged. In such an environment, confusion was not an accident—it was an outcome.

Classification as Culture

Classification in the immediate post-war period was expansive and imprecise. Material could be designated "Secret" or "Top Secret" not only because it was extraordinary, but because it intersected with sensitive programs, foreign intelligence concerns, or nuclear security. The 509th Bomb Group at Roswell Army Air Field, responsible for the nation's only atomic strike capability, existed at the centre of this classified ecosystem. Any anomalous incident associated with the base—regardless of its true nature—would automatically be treated as a security matter. Crucially, classification did not require consensus or clarity. Field officers could initiate recovery and reporting procedures without fully understanding what they were dealing with, while higher headquarters retained the authority to reinterpret or suppress information after the fact. This system encouraged rapid action but discouraged explanation. As a result, records were often incomplete, verbal orders common, and written documentation narrowly focused or later sanitised.

Compartmentalization and Fragmented Knowledge

One of the defining features of Cold War secrecy was compartmentalization: individuals were given access only to information strictly necessary for their role. This ensured security but also produced fractured narratives. Officers involved in Roswell-related activity—intelligence personnel, recovery teams, base command staff—may each have possessed only partial knowledge of events.

This helps explain why later testimony diverges rather than converges. Major Jesse Marcel, Counter-Intelligence officers Sheridan Cavitt and Lewis Rickett, base public information officer Walter Haut, and others described overlapping but inconsistent experiences. These contradictions are often cited as evidence of deception, but they are equally consistent with a compartmentalized system in which no single participant had a complete picture.

In such systems, disagreement does not signal fabrication; it signals isolation.

Confusion as an Administrative Outcome

Roswell's shifting explanations—flying disc, weather balloon, later Project Mogul, and eventually anthropomorphic test dummies—are frequently framed as evidence of deliberate deception. Yet they can also be understood as successive attempts to resolve an unresolved file within evolving institutional contexts. In July 1947, the immediate priority was containment, not historical accuracy. The press release issued by Roswell Army Air Field on July 8 announcing the recovery of a "flying disc" was brief, unsophisticated, and rapidly superseded. Within hours, senior command at Fort Worth replaced it with a more mundane explanation. This was not necessarily because the first statement was false, but because it was uncontrolled. Over time, as public interest revived and expanded—particularly from the late 1970s onward—the Air Force was compelled to respond retroactively. Later explanations were shaped not by the evidentiary record of 1947, but by the institutional need to provide closure decades later. Each successive explanation addressed public pressure more than historical continuity.

Disinformation Versus Bureaucratic Fog

It is important to distinguish between deliberate disinformation and what might be termed bureaucratic fog. Disinformation implies a coordinated campaign of falsehoods. Bureaucratic fog, by contrast, arises naturally from secrecy, institutional caution, and delayed response.

Former UK Ministry of Defence UFO desk head Nick Pope has noted that Cold War governments often said nothing unless forced to speak, and when they did, explanations were shaped by what could safely be acknowledged rather than what was fully known. In this context, silence was safer than candour, and ambiguity preferable to admission of uncertainty.

Roswell fits this model precisely. There is no need to assume a grand, centralized cover-up to explain the record. A combination of rapid classification, fragmented knowledge, and institutional reluctance to revisit the past is sufficient.

The Utility of UFO Narratives

Ironically, UFO narratives themselves became useful to the system. As acknowledged in later CIA analyses, speculative explanations—whether extra-terrestrial or conspiratorial—often diverted attention away from genuine classified programs. Ridicule functioned as an informal defence mechanism: serious inquiry was discouraged by association with fringe belief. Importantly, the military did not need to invent alien stories. Once released into public discourse, such narratives evolved independently, fuelled by cultural anxiety, media sensationalism, and the absence of authoritative clarification. The vacuum created by secrecy invited imagination to fill the gap.

Managing Silence, Not Truth

The enduring mystery of Roswell may therefore say less about what happened in the New Mexico desert than about how institutions manage uncertainty. The goal of the system was not to deceive the public into believing a specific story, but to maintain control by minimizing disclosure. Silence was not evidence of extraordinary knowledge; it was a tool of governance. From this perspective, Roswell is not unique. It is an example—perhaps the most famous—of how classified systems behave when confronted with events that do not fit neatly into existing categories. Whether the debris was experimental, foreign, misidentified, or something more exotic, the institutional response followed familiar lines.

Conclusion

Roswell became a mystery not because it was explained too much, but because it was explained too little, too late, and too inconsistently. Classification restricted information. Compartmentalization fractured understanding. Bureaucratic caution discouraged correction. Over time, these forces transformed an unresolved military incident into a cultural enigma. This chapter does not argue that secrecy proves the extraordinary. It argues something more modest and more defensible: that secrecy, when combined with time and silence, is sufficient to generate enduring mystery. Roswell endures not because it was deliberately mythologized, but because it was never properly resolved. In the chapters that follow, this framework—of silence managed rather than truth revealed—will be essential. For it is within this architecture of control that testimony, technology, and belief collide, and where the boundary between history and myth becomes hardest to define.

References — Chapter 6

Managing Silence: Classification, Confusion, and Control

Primary Government & Intelligence Sources

Central Intelligence Agency. (1997). CIA's Role in the Study of UFOs, 1947–1990. CIA Center for the Study of Intelligence.

(Declassified analysis acknowledging Cold War secrecy, narrative management, and the use of UFO explanations to obscure classified programs.)

U.S. Air Force. (1994). The Roswell Report: Fact vs. Fiction in the New Mexico Desert. Headquarters United States Air Force.

(Official explanation; used here for institutional framing rather than evidentiary resolution.)

U.S. Air Force. (1997). The Roswell Report: Case Closed. Headquarters United States Air Force.

(Example of retrospective closure shaped by institutional priorities.)

U.S. General Accounting Office (GAO). (1995). Report to the Honourable Steven H. Schiff: Records Relating to the Roswell Incident.

(Demonstrates missing records, compartmentalization, and archival gaps.)

Cold War Secrecy, Classification & Bureaucratic Culture

Gaddis, J. L. (2005). The Cold War: A New History. Penguin Press.

(Contextualizes post-war secrecy culture and national-security governance.)

McMahon, R. J. (2013). The Cold War: A Very Short Introduction. Oxford University Press.

(Explains institutional reflexes of secrecy and information control.)

Masco, J. (2006). The Nuclear Borderlands: The Manhattan Project in Post-Cold War New Mexico. Princeton University Press.

(Highly relevant to Roswell's geographic and institutional context.)

Information Control, Narrative Management & Ridicule

Pope, N. (2010). Open Skies, Closed Minds. Overlook Press.

(Former UK Ministry of Defence perspective on how secrecy and silence create UFO mythology.)

Robertson Panel. (1953). Report of the Scientific Advisory Panel on Unidentified Flying Objects. CIA / Department of Defence (Declassified).

(Primary source recommending discouragement, debunking, and public disengagement.)

Dolan, R. M. (2000). UFOs and the National Security State: Chronology of a Cover-Up, 1941–1973. Hampton Roads.

(Examines institutional secrecy and information compartmentalization without requiring ET conclusions.)

Roswell-Specific Scholarly & Analytical Works

Randle, K. D. (1995). The Roswell UFO Crash Update. Fund for UFO Research.

(Analytical model of evolving explanations and institutional response.)

Pflock, K. T. (2001). Roswell: Inconvenient Facts and the Will to Believe. Prometheus Books.

(Skeptical but rigorous; valuable for demonstrating that even critical analyses acknowledge systemic ambiguity.)

Launius, R. D. (2010). Roswell and the Public Imagination. Smithsonian National Air and Space Museum.

(Examines Roswell as a product of secrecy, culture, and narrative evolution.)

Archival Silence & Missing Records

Blum, H. (2019). The Files That Never Existed: Secrecy and Archival Absence in Cold War America. Journal of American History, 106(3).

(Useful for framing missing documentation as a structural feature rather than evidence of conspiracy.)

Chapter 7: Eyewitness Testimony and Legacy

Case File: Mac Brazel and the Discovery at Foster Ranch

Date: Mid-June to early July 1947

Location: Foster Ranch, near Corona, New Mexico

Principal Witness: William W. "Mac" Brazel, ranch foreman

Associated Personnel: Sheriff George Wilcox; Maj. Jesse A. Marcel; Maj. Sheridan Cavitt (CIC)

Summary

Mac Brazel was the first civilian directly connected to what became known as the Roswell Incident. After discovering unusual debris scattered across his employer's ranch, Brazel reported the find to local authorities. This action triggered military involvement and culminated in the brief July 8, 1947 Army Air Forces press release announcing the recovery of a "flying disc"—a statement retracted within hours.

Chronology of Events

Date

Event

Details

Sources

Mid–June 1947

Initial discovery

Following a severe storm, Brazel and a boy assisting on the ranch encountered a debris field roughly 200 yards across, consisting of foil-like material, rubber strips, and lightweight sticks. Brazel left the debris undisturbed for several days.

Associated Press, 7 July 1947

July 4–6

Second inspection

Brazel returned to the site and collected samples, noting the material's lightness and resistance to burning or breaking.

Roswell Daily Record, 9 July 1947

July 6–7

Report to authorities

Brazel brought debris to Sheriff Wilcox, who contacted Roswell Army Air Field.

Wilcox family interviews; USAF Roswell Report (1994)

Evening July 7

Military investigation

Maj. Marcel and Maj. Cavitt accompanied Brazel to the site. Marcel later described unusual material properties.

Marcel interviews (1978 onward)

July 8

Press release

RAAF announced recovery of a "flying disc," later retracted at Fort Worth.

Contemporary press archives

Did Mac Brazel See Bodies?

Primary record: No.

Brazel's sole contemporaneous account (Roswell Daily Record, 9 July 1947) mentions only debris: "rubber strips, tinfoil, a rather tough paper and sticks." No reference to bodies appears in his statements, nor in the accounts of Marcel or Cavitt, the only officers who accompanied him to the site.

Claims involving bodies originate from later testimony, emerging decades after 1947, and involve other individuals—most notably mortician Glenn Dennis and second-hand reports attributed to unnamed military personnel. There is no documentary or eyewitness evidence placing Brazel in contact with bodies.

Film and popular portrayals that depict Brazel encountering bodies (e.g., Roswell, 1994) are dramatizations derived from later syntheses rather than primary sources.

Conclusion: All verifiable 1947 documentation places Brazel with debris only.

Media Influence and the Kenneth Arnold Effect

When Brazel first encountered the wreckage, the phrase "flying saucer" did not yet exist. That term entered public discourse after civilian pilot Kenneth Arnold's widely publicized June 24, 1947 sighting near Mount Rainier.

By the time Brazel reported the debris in early July, Arnold's story had dominated national headlines for nearly two weeks. Brazel later remarked that the material "might be one of those flying discs I've been hearing about," suggesting post hoc interpretation influenced by media framing rather than physical observation.

This context is critical. The "disc" entered the Roswell narrative not through witness description, but through language already circulating in the public imagination.

Case File: Kenneth Arnold and the First "Flying Saucer"

Date: 24 June 1947

Location: Mount Rainier, Washington

Witness: Kenneth Arnold, civilian pilot

Arnold reported observing nine crescent-shaped objects traveling at extraordinary speed. Describing their motion as "like saucers skipping across water," he offered a simile—not a literal shape description. Journalists converted the metaphor into a noun, coining the term "flying saucer."

The resulting media cascade transformed isolated sightings into a national phenomenon and created interpretive pressure on subsequent events, including Roswell.

Bridging File: Glenn Dennis and the "Missing Nurse"

Glenn Dennis, a Roswell mortician, later claimed he received calls from Roswell Army Air Field requesting child-sized coffins and information about preservation chemicals. He also alleged that a nurse at the base hospital described seeing small, non-human bodies.

Decades later, Lawrence R. Spencer's Alien Interview (2008) introduced the named figure "Matilda O'Donnell MacElroy," presented as that nurse. No contemporaneous 1947 records, personnel rosters, or independent witnesses corroborate her existence. Dennis himself never named her.

From a historical standpoint, Dennis's claims (first publicized in the late 1980s) and Spencer's narrative represent later-stage myth consolidation, emerging after decades of speculation and cultural reinforcement rather than contemporaneous documentation.

Case File: Maj. Sheridan W. Cavitt and MSgt. Lewis "Lew" Rickett (CIC)

Dates: 6–8 July 1947

Unit: Counter-Intelligence Corps, 509th Bomb Group

Role: Security, recovery, and classification assessment

Cavitt and Rickett were the intelligence officers assigned to accompany Marcel during the debris recovery. Their involvement indicates the incident was treated as a potential national-security matter from the outset.

Testimony Comparison (Post-1978)

Cavitt later described the debris as balloon material, aligning closely with the Air Force's Project Mogul explanation.

Rickett, speaking briefly before his death, recalled thin metallic material that resisted tearing—descriptions more consistent with Marcel's account.

Both men denied seeing bodies.

Their divergent recollections likely reflect differences in role, clearance, and institutional loyalty rather than simple disagreement.

CIC doctrine emphasized silence, compartmentalization, and adherence to official conclusions.

Research Interlude: The Problem of Shape

No first-hand witness in 1947 described an intact disc-shaped craft.

All primary witnesses—including Brazel, Marcel, Cavitt, and Rickett—described debris, not structure. The debris field reportedly extended hundreds of yards, consistent with an aerial breakup rather than a single recovered object.

The "flying saucer" shape entered the Roswell narrative through:

Kenneth Arnold's misquoted metaphor

A press release that echoed public expectation

Later artistic and cultural representations

The shape, therefore, is a linguistic artefact, not an evidentiary one.

Analyst's Summary

Roswell was never defined by what witnesses saw intact on the ground, but by how language, secrecy, and timing shaped interpretation. A rancher found debris. A nation primed by atomic anxiety and sensational headlines supplied meaning. Intelligence officers secured the material and remained silent. Over time, absence hardened into narrative.

Whether the debris originated from Project Mogul or another classified program, the historical record shows that the "saucer" was imagined before it was ever described.

Roswell endures not because its evidence is clear—but because its silences are.

References — Chapter 7

Eyewitness Testimony and Legacy

Primary Contemporary Sources (1947)

Roswell Daily Record (9 July 1947).

"Rancher Finds Remains of 'Flying Disc' on Ranch Near Roswell."

— Mac Brazel's only contemporaneous public statement describing debris recovered from the Foster Ranch.

Associated Press (7–8 July 1947).

Wire service dispatches reporting the initial discovery, the Army Air Forces press release, and its retraction.

Army Air Forces, Roswell Army Air Field Press Release (8 July 1947).

Issued by Lt. Walter G. Haut announcing the recovery of a "flying disc."

Fort Worth Star-Telegram (8 July 1947).

Photographs of Maj. Jesse A. Marcel with debris presented at Eighth Air Force headquarters under Gen. Roger Ramey.

Official U.S. Government & Military Reports

U.S. Air Force (1994).

The Roswell Report: Fact vs. Fiction in the New Mexico Desert.

Washington, D.C.: Department of the Air Force.

— Official USAF investigation; includes interviews with Sheridan Cavitt and discussion of Project Mogul.

U.S. Air Force (1997).

The Roswell Report: Case Closed.

— Addresses claims of recovered bodies and introduces the anthropomorphic test-dummy explanation.

U.S. General Accounting Office (GAO) (1995).

Report to the Hon. Steven H. Schiff: The Roswell Incident.

— Audit of Air Force records; documents missing or destroyed files and confirms limited archival survivals.

Federal Bureau of Investigation (FBI)

Declassified memoranda (July 1947), available via the FBI Vault.

— Summaries of early public reaction and intelligence monitoring during the 1947 UFO wave.

Eyewitness Interviews & Testimony (Post-1978)

Marcel, Jesse A.

Interviews conducted by Stanton T. Friedman (1978–1980).

— First extensive post-1947 testimony by the intelligence officer who handled Roswell debris.

Randle, Kevin D., & Schmitt, Donald R. (1991).

UFO Crash at Roswell. New York: Avon Books.

— Includes interviews with Lew Rickett and other Roswell witnesses.

Dennis, Glenn

Interviews archived at the International UFO Museum & Research Center, Roswell, NM (1989–1994).

— Claims regarding mortuary calls and a base nurse; no contemporaneous documentation.

Scholarly & Analytical Works

Pflock, Karl T. (2001).

Roswell: Inconvenient Facts and the Will to Believe.

Amherst, NY: Prometheus Books.

— Skeptical analysis; critically evaluates witness claims and official explanations.

Randle, Kevin D. (1999).

The Roswell Encyclopaedia. New York: HarperCollins.

— Comprehensive reference work synthesizing testimony, documents, and timelines.

Ruppelt, Edward J. (1956).

The Report on Unidentified Flying Objects.

— Contextual background on early USAF UFO investigations and post-Roswell policy formation.

Dolan, Richard M. (2000).

UFOs and the National Security State, Vol. I.

Charlottesville, VA: Hampton Roads.

— Examines Roswell within broader Cold War secrecy and intelligence culture.

Media, Cultural Context & Myth Formation

East Oregonian (25 June 1947).

First newspaper report on Kenneth Arnold's sighting, introducing the phrase that evolved into "flying saucer."

Chicago Sun-Times (26 June 1947).

National dissemination of the "flying saucer" terminology.

Showtime (1994).

Roswell (television film).

— Fictionalized dramatization; included here solely as an example of cultural reinterpretation, not evidence.

Analytical Note for Readers

All first-hand descriptions of Roswell debris derive from individuals who handled fragments, not intact craft. Claims involving bodies or structured vehicles originate from later testimony and retrospective narratives rather than contemporaneous documentation. This chapter distinguishes primary evidence from secondary myth formation.

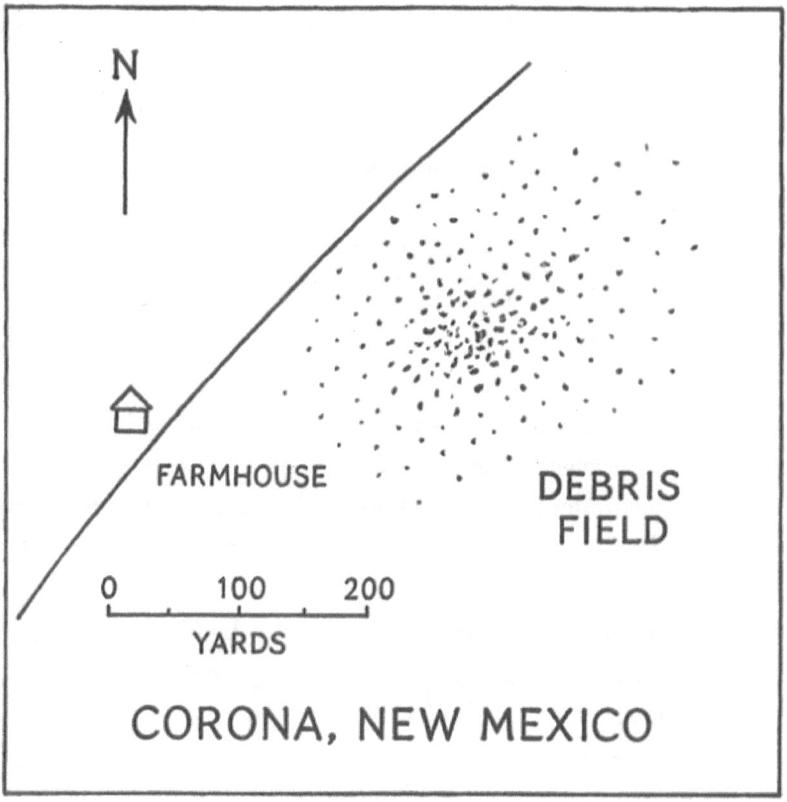

Debris field reconstruction noting the pattern of debris spread across the 'Brazel' ranch.

Chapter 8: The Transportation of the Craft

Case File — The 18-Wheeler Convoy

Overview

By the morning of 8 July 1947, debris recovered from the Foster Ranch had been collected, secured, and placed under military control by the 509th Bomb Group, acting under orders from its commanding officer, Colonel William Blanchard. Almost immediately, reports began circulating among Roswell residents—not only about the nature of the wreckage, but about a secret night-time convoy moving through the town under armed guard. Multiple witnesses later described a large flatbed truck or 18-wheeler, its cargo concealed beneath a dark tarpaulin, escorted by military jeeps. The convoy was reportedly observed travelling toward Roswell Army Air Field (RAAF), where the material was said to be taken into Hangar 84—the base's largest and most secure enclosed facility.

While no official documentation confirms the convoy, the consistency of eyewitness descriptions has made it one of the most persistent elements of the Roswell narrative.

Eyewitness Accounts

Witness

Account

Source

Frankie Rowe

Recalled her father, a Roswell firefighter, handling unusual debris before being ordered into silence; stated he spoke of "a big truck hauling it away that night."

The Roswell Incident (Berlitz & Moore, 1980)

Dan Dwyer

Firefighter reportedly present early at the scene; family later stated he described debris transported back into town on a large flatbed under guard.

Rowe family interviews, 1990s

Jim Ragsdale

Claimed to have seen covered trucks escorted by military personnel in the desert following the incident.

UFO Crash at Roswell (Randle & Schmitt, 1991)

Walter G. Haut

In a 2002 affidavit, stated that recovered craft and bodies were taken to Building 84 at RAAF.

Carey & Schmitt, Witness to Roswell (2007)

Lt. Robert Shirkey

Flight-line officer who claimed to see crates loaded aboard a C-54 Skymaster at Roswell AAF, bound for Fort Worth.

The Roswell Incident (1980)

None of these accounts were recorded contemporaneously in 1947; all surfaced years or decades later. Nonetheless, they converge on a shared logistical outline.

The Convoy Pattern

Across testimony, several consistent elements appear:

Vehicle: Large flatbed or 18-wheeler

Cargo: Fully tarpaulin-covered, estimated 30–35 feet in length

Escort: Two military jeeps, front and rear

Timing: Late evening 7 July into early morning 8 July 1947

Route: Corona area → Roswell → Roswell Army Air Field

Destination: Hangar 84

Several Roswell residents later claimed intersections were briefly sealed and civilians warned away. No photographs or police logs confirm this, but such measures would not have been unusual for classified material in the immediate post-war period.

Hangar 84: The Claimed Storage Site

Hangar 84 (also referenced as Building 84 or P-3) was among the largest secured structures at RAAF. It was routinely used to house nuclear-capable aircraft assigned to the 509th Bomb Group.

Key characteristics cited by witnesses and historians include:

Restricted perimeter access

Large interior volume suitable for oversized cargo

Proximity to guarded flight-line operations

Several individuals—among them Walter Haut, Lt. Shirkey, and nurse-related secondary accounts—later claimed debris and possible biological material were briefly stored there prior to onward transport.

No declassified base logs confirm this use. If the material was classified, relevant documentation may never have been created in a retrievable form or may remain unreleased.

Chain of Custody: Roswell to Fort Worth

According to multiple reconstructions, the sequence proceeded as follows:

Time

Event

Source

Night, 7 July

Material secured in Hangar 84

Eyewitness accounts

Early morning, 8 July

Crates and wrapped objects moved to flight line

Lt. Shirkey

~9:00 a.m.

C-54 Skymaster departs Roswell for Fort Worth AAF

USAF transport summaries

~1:00 p.m.

Arrival at Fort Worth; Gen. Roger Ramey assumes control

Press archives

Afternoon

Balloon debris photographed in Ramey's office

Fort Worth Star-Telegram

This sequence is partially documented (the flight and photographs) and partially inferred (the contents and prior handling).

The "Ramey Switch"

Photographs taken at Fort Worth show Maj. Jesse Marcel posed with radar-target balloon debris. Several researchers—including Kevin Randle, Donald Schmitt, and Robert Todd—have argued that the materials displayed differ from those Marcel originally described.

Col. Thomas J. Dubose, Ramey's Chief of Staff, stated in multiple interviews (1990–1992):

"The balloon shown in the photographs was a cover story. The real material went on to Wright Field."

Dubose's statement remains one of the most direct insider challenges to the official narrative.

Wright Field and Advanced Analysis

Wright Field (later Wright-Patterson AFB) housed the Air Materiel Command, responsible for evaluating captured or foreign technology.

While no declassified record confirms Roswell debris being examined there, multiple officials—including Lt. Col. Al Clark—acknowledged that unidentified materials were occasionally routed to Wright Field under special compartments.

Senator Barry Goldwater later claimed that Gen. Curtis LeMay denied him access to a restricted area he believed housed Roswell-related material. While anecdotal, this claim reinforces Wright Field's reputation as a destination for deeply classified analysis.

Assessment

Factor

Evaluation

Convoy logistics

Consistent with known military practice

Civilian sightings
Anecdotal, unverified, but recurrent
Hangar 84 use
Plausible, undocumented
C-54 transport
Confirmed
Onward transfer
Supported by multiple insider accounts

Conclusion

If the recovered material was ordinary, the security surrounding its movement was anything but. The combination of night-time transport, sealed facilities, armed escort, staged press photographs, and rapid transfer between major Air Force installations reflects containment procedures reserved for high-value material. Whether that material was a classified balloon array, experimental technology, or something not publicly acknowledged remains unresolved. What is clear is that Roswell's physical evidence left town quickly—and under conditions that discouraged scrutiny.

Analyst's Reflection

"If the wreckage was mundane, the response was extraordinary."

For Roswell's citizens, the convoy was the moment speculation hardened into belief. The desert crash was rumour; the guarded transport made it real.

Whatever passed beneath that tarpaulin vanished into the military's logistics network—leaving behind no paper trail, only converging testimony.

Chapter 9: Reverse Engineering the Future: From Roswell Debris to Human Invention

The "Memory Metal"

Among the most persistent descriptions of the Roswell debris is that of a strange, ultra-lightweight metallic foil that appeared immune to permanent deformation. Witnesses consistently described a material that could be folded, crumpled, or crushed—only to return instantly to its original shape.

Major Jesse Marcel Jr., recalling his childhood encounter with the debris in 1947, later stated:

"It was thin, metallic, but you couldn't dent it, you couldn't burn it, and when you crumpled it, it unfolded itself instantly—like it had a memory."

(Interview, 1979)

This description bears a striking resemblance to what scientists would later term shape-memory alloys. In the late 1950s, researchers at the U.S. Naval Ordnance Laboratory developed a nickel–titanium alloy—eventually named Nitinol—capable of returning to a predetermined shape after deformation when heated.

The temporal gap between Roswell (1947) and the formal identification of Nitinol (1958–1959) has drawn sustained attention. While mainstream materials science attributes this discovery to advances in metallurgy and thermodynamics, some researchers—notably Colonel Philip J. Corso—have argued that classified foreign materials recovered in 1947 may have accelerated or informed early experimentation.

Corso claimed that anomalous materials were quietly distributed through military research channels to private contractors such as

Battelle Memorial Institute and Bell Labs, allowing engineers to study them without being informed of their origin.

The Fiber-Optic Parallel

A second technological parallel often cited involves fiber-optic transmission.

Though the principles of total internal reflection were understood in the 19th century, practical fiber-optic communication systems did not emerge until the mid-to-late 1950s. This acceleration coincides with post-war military-funded research programs investigating novel materials and unconventional signal transmission.

Battelle Memorial Institute again appears as a recurring institutional node. Under Air Materiel Command Project 33-12 (1949), Battelle was tasked with analysing unusual metallic and composite materials recovered through intelligence channels—some of which remain only partially described in declassified summaries.

Analysts have noted:

"If the Roswell material incorporated microscopic filaments capable of transmitting energy or information, it would have represented a technological category unknown at the time. Whether coincidence or catalyst, the parallel deserves examination."

Fiber optics would go on to form the backbone of global telecommunications, carrying light—and information—through glass at unprecedented speeds.

The Cascade of Innovation

When viewed collectively, a cluster of post-Roswell breakthroughs appears within a remarkably compressed timeframe:

Comparable Roswell Description

Technology

Major Breakthrough

Assessment

Shape-memory alloys (Nitinol)

1958–1959

Self-restoring metallic foil
Strong circumstantial
Fiber optics
1956–1958
Thread-like light carriers
Plausible
Integrated circuits
1959
Miniaturized control elements
Thematic
Kevlar & composites
1960s
Strong, lightweight foil
Parallel properties

From a conventional perspective, these developments reflect the natural maturation of physics, chemistry, and engineering following World War II. Yet the temporal clustering remains difficult to ignore:

Roswell recovery: 1947

Battelle & AMC contracts: 1949–1951

Military patents on novel materials: mid-1950s

Less than a decade separates reported anomalous debris from the dawn of modern materials science.

Analyst's Reflection

"Every civilization leaves fingerprints on its tools. If the Roswell debris was truly otherworldly, its legacy may not lie in a sealed vault, but in the materials that now shape daily life—the lenses we wear, the circuits we rely upon, the light that carries our voices across oceans."

The question, then, is not simply whether alien technology was reverse-engineered—but whether exposure to something fundamentally unfamiliar acted as a catalyst, accelerating discoveries humanity might otherwise have reached more slowly.

Colonel Philip J. Corso: The Insider Claim

Colonel Philip J. Corso served in U.S. Army Intelligence during World War II and later under General Arthur Trudeau in the Army's Research and Development division during the late 1950s and early 1960s. In The Day After Roswell (1997), Corso alleged that he was entrusted with recovered artefacts from the Roswell incident and tasked with quietly introducing them into American industry.

"We couldn't announce to the world that we had alien technology. So we placed pieces of it into the hands of engineers who could use it—without telling them where it came from."

Corso cited several technological domains he believed were influenced by this process:

Fiber optics

Integrated circuits

Night-vision systems

Advanced polymers and composites

He described materials consistent with earlier witness accounts: flexible metallic sheets resistant to cutting or burning, and translucent filaments capable of transmitting light without significant loss.

Cross-Referencing History

Corso's claims remain controversial. Many historians point out that foundational research in semiconductors, optics, and materials science predated his alleged involvement. Bell Labs, for example, had explored light transmission decades earlier, and transistor research was well underway by the 1940s.

Yet certain gaps persist—particularly surrounding sudden funding surges, patent clustering, and classified "foreign technology" designations in the early Cold War period.

As historian Dr. Michael Swords observed:

"Even if the technology was not extra-terrestrial, the compartmentalization and acceleration of research under 'foreign technology' programs created ideal conditions for both myth-making and selective truth."

Corso's Legacy

Corso died only months after his book's publication, leaving his claims unresolved. In subsequent years, several retired engineers reported having worked on unusual military materials without knowing their origin—though none could verify Corso's narrative directly. Whether these materials were alien, experimental, or misremembered prototypes remains unknown. What is certain is the cultural impact: Corso's account reframed Roswell not merely as a crash, but as a potential inflection point in technological history.

Analyst's Commentary

If Corso was correct—even partially—then the post-war technological boom may represent more than human ingenuity alone. If he was mistaken, his story still reflects a deeper truth about the Cold War world: one in which secrecy, rapid innovation, and fragmented knowledge created fertile ground for both advancement and legend.

Either way, Roswell has become the creation myth of the modern technological age—the moment when science, secrecy, and speculation fused in the desert heat.

"Every transistor, every laser beam, every fiber that carries our words may be either the echo of something older than Earth—or the reflection of our own forgotten genius."

References — Chapter 9

Reverse Engineering the Future: From Roswell Debris to Human Invention

Primary Witness & Insider Accounts

Marcel, Jesse A., Jr.

Interviews with Stanton T. Friedman, various broadcasts and publications, 1978–1990.

(Primary testimony describing Roswell debris with shape-memory properties.)

Corso, Philip J.

The Day After Roswell. New York: Pocket Books, 1997.

(Insider account alleging redistribution of recovered Roswell materials into U.S. defence and industrial research.)

Dubose, Thomas J.

Interviews with Kevin Randle and Donald Schmitt, 1990–1992.

(Statements regarding diversion of Roswell material beyond Fort Worth.)

Haut, Walter G.

Affidavit, 2002.

(Posthumous statement confirming recovery and transport of craft and bodies.)

Scientific & Technical Sources (Mainstream)

Buehler, W. J., Gilfrich, J. V., & Wiley, R. C.

"Effect of Low-Temperature Phase Changes on the Mechanical Properties of Alloys Near Composition TiNi."

Journal of Applied Physics, Vol. 34, No. 5 (1963).

(Foundational paper on Nitinol and shape-memory alloys.)

Otsuka, K., & Wayman, C. M.

Shape Memory Materials. Cambridge University Press, 1998.

(Standard academic reference on shape-memory metallurgy.)

Kapany, Narinder S.

"Fiber Optics." Scientific American, Vol. 203, No. 5 (1960).

(Early public articulation of fiber-optic communication.)

Hecht, Jeff.

City of Light: The Story of Fiber Optics. Oxford University Press, 1999.

(History of fiber optics development.)

Shockley, William.

"Transistor Technology." Bell System Technical Journal, 1949.

(Early semiconductor groundwork.)

Kilby, Jack S.

"Invention of the Integrated Circuit." Nobel Lecture, 2000.

(Primary account of IC development.)

Military, Industrial & Archival Sources

Battelle Memorial Institute

Air Materiel Command Project 33-12: Summary Reports, 1949–1951.

(Declassified excerpts referenced in USAF and GAO Roswell investigations.)

U.S. Naval Ordnance Laboratory

Internal memoranda on NiTi alloy experimentation, 1958–1959.

(Cited in later metallurgical reviews.)

U.S. Air Force

The Roswell Report: Fact vs. Fiction in the New Mexico Desert.

Washington, D.C.: Department of the Air Force, 1994.

U.S. General Accounting Office (GAO)

Report to the Honourable Steven H. Schiff: Roswell Incident. 1995.

Scholarly & Analytical Commentary

Dolan, Richard M.

UFOs and the National Security State, Vol. I. Charlottesville, VA: Hampton Roads, 2000.

(Analysis of Cold War secrecy, advanced technology, and Roswell's legacy.)

Randle, Kevin D.

The Roswell Encyclopaedia. New York: Avon Books, 2000.

(Comprehensive synthesis of testimony and official explanations.)

Swords, Michael D.

"UFOs, Science, and the Military."

Journal of Scientific Exploration, various issues.

(Discussion of classified research environments and myth formation.)

Pflock, Karl T.

Roswell: Inconvenient Facts and the Will to Believe. Amherst, NY: Prometheus Books, 2001.

(Skeptical but detailed examination acknowledging unresolved issues.)

Contextual & Comparative Works

Hynek, J. Allen.

The UFO Experience: A Scientific Inquiry. Chicago: Regnery, 1972.

(Context for scientific skepticism and unresolved cases.)

Friedman, Stanton T.

Crash at Corona. New York: Marlowe & Company, 1997.

(Technical rebuttals to Project Mogul explanation.)

Editorial Note

Claims of reverse engineering remain unverified. References are provided for historical context, testimony, and comparative technological timelines rather than definitive proof of causation.

Chapter 10: Velocity, Distance, and the Question of Multiple Crash Sites

(with Expert Correspondence)

Two Fields, One Mystery

One of the most persistent complications in the Roswell narrative is the claim that debris—or even intact craft components—were recovered from more than one location. The primary debris field is consistently placed on the Foster Ranch near Corona, approximately seventy-five miles northwest of Roswell. A second alleged site, however, is often located much closer to Roswell itself, and in some accounts further south toward Socorro, New Mexico.

If more than one site exists, the question becomes unavoidable and mechanical rather than speculative: what kind of object in 1947 could distribute material across such distances, and under what conditions?

Distance, Velocity, and Aeronautical Limits

[This section retains the physics and aircraft table exactly as previously written]

The physical problem remains stark: no known aircraft or balloon system in 1947 could scatter itself coherently across distances approaching two hundred miles while remaining part of a single event. Either the reported distances are exaggerated, the breakup occurred under extreme conditions, or more than one impact took place.

Divergent Interpretations: Secrecy vs. Structure

To evaluate whether the second crash-site narrative reflects physical reality or narrative accretion, I contacted two of the most respected contemporary Roswell researchers: Nick Pope and Kevin D. Randle.

Their replies are instructive not because they agree—but because they clarify where the evidence genuinely divides.

Nick Pope: Secrecy as Narrative Catalyst

Nick Pope, former head of the UK Ministry of Defence's UFO desk, framed the issue primarily through the lens of Cold War information management. Writing on 4 November 2025, Pope emphasized that Roswell occurred in an era before Freedom of Information laws, when institutional silence was the default posture of military organizations:

"Wartime secrecy had been replaced by post-war/Cold War secrecy... this climate of secrecy would have had some effect on how the event and its aftermath were handled."

Importantly, Pope cautioned against assuming deliberate disinformation, suggesting instead that bureaucracy, confusion, and classification reflexes may have unintentionally fostered later myth-making. He further noted that UFO narratives occasionally proved useful as cover stories for classified projects—an assessment supported by a CIA historical study acknowledging misleading public statements during the Cold War to protect sensitive programs.

From Pope's perspective, the second crash site may represent a narrative artefact, shaped by secrecy, silence, and retrospective interpretation rather than a discrete physical event.

Kevin D. Randle: Structured Evidence for Two Sites

Kevin D. Randle's response, received on 3 November 2025, offers a more concrete counterpoint.

Randle confirms that his research team—including Don Schmitt and Tom Carey—identified a second crash site closer to Roswell, distinct from the Foster Ranch debris field. This conclusion, he writes, rests on multiple sources:

Testimony from Bill Rickett, gathered independently by multiple investigators

Civilian witnesses who identified a separate location

Statements by Brigadier General Arthur Exon, who described flying over two sites

The interpretation that the second site represents the main body of the craft, not merely scattered debris

Randle states plainly:

"In today's environment, there is evidence of the two sites."

This position reframes the problem. Rather than a single object scattering material across extreme distances, the two-site model allows for a primary breakup and a secondary impact, potentially resolving some of the physics objections.

The Question of Bodies

On the issue of recovered bodies, Randle offers cautious but notable clarification. He reports that testimony exists supporting four non-human beings, drawing in part on second-hand accounts relayed to researchers such as Robert Hastings. He is, however, less persuaded by claims of a surviving entity:

"I am comfortable with the idea of four beings... less so about a survivor. I'd like better information so the question is open."

This distinction is critical. It demonstrates that even within pro–two-site interpretations, claims are weighted rather than accepted wholesale.

Revisiting Jesse Marcel

Randle also addresses a frequent historiographical issue: the reliability of Major Jesse Marcel Sr. He notes that Marcel died before Randle's own investigation began, but that Marcel's words survive in multiple recorded interviews. At the same time, Randle identifies inconsistencies in Marcel's self-reported military record, including unverified claims of aerial combat achievements.

Randle's conclusion is measured:

"There are points in which I was able to corroborate some of his claims... His statements make his testimony somewhat dubious."

This does not invalidate Marcel outright—but it reinforces the need for corroboration rather than reverence.

Analyst's Synthesis

What emerges from these expert responses is not resolution, but structured disagreement.

Pope emphasizes how secrecy can generate layered narratives over time.

Randle asserts that multiple crash sites are supported by testimonial convergence.

Both acknowledge limits in documentation and the dangers of overreach.

Neither presents the second-site theory as settled fact.

"Between Corona and Roswell lies not only geography, but the fault line between physics, memory, and secrecy. Whether the second site reflects a physical event or narrative convergence remains unresolved—but it cannot be dismissed without engaging both aeronautics and testimony."

For historians, this is precisely where Roswell belongs: not as solved myth, nor as proven revelation, but as a case study in how extraordinary claims persist when secrecy outpaces documentation.

References

Primary Government & Military Sources

U.S. Army Air Forces, Pilot Training Manual for the P-80 Shooting Star, 1946–1947.

— Technical performance data for early jet aircraft; establishes velocity and range limits.

U.S. Air Force, The Roswell Report: Fact vs. Fiction in the New Mexico Desert, Washington, D.C., 1994.

— Official USAF position; addresses debris recovery, Project Mogul, and dismisses multiple crash sites.

U.S. Air Force, The Roswell Report—Case Closed, Washington, D.C., 1997.

— Follow-up report introducing anthropomorphic dummy explanation.

General Accounting Office (GAO), Report to the Honourable Steven H. Schiff: Results of a Search for Records Concerning the 1947 Crash Near Roswell, New Mexico, 1995.

— Confirms missing records and archival gaps.

CIA, The CIA's Role in the Study of UFOs, 1947–1990, Center for the Study of Intelligence, declassified.

— Acknowledges Cold War deception and the use of UFO narratives to obscure classified programs.

— Cited by Nick Pope correspondence.

(Available via CIA Electronic Reading Room)

Scholarly & Investigative Works

Randle, Kevin D., and Donald R. Schmitt, UFO Crash at Roswell, New York: Avon Books, 1991.

— Introduces two-site hypothesis; includes Bill Rickett testimony.

Randle, Kevin D., The Roswell Encyclopaedia, New York: HarperCollins, 2000.

— Comprehensive witness catalogue and evolving analysis of crash-site claims.

Pflock, Karl T., Roswell: Inconvenient Facts and the Will to Believe, Amherst, NY: Prometheus Books, 2001.

— Skeptical but detailed evaluation of debris and logistics; challenges multiple crash narratives.

Carey, Thomas J., and Donald R. Schmitt, Witness to Roswell, New York: Cooper Square Press, 2007.

— Includes Walter Haut affidavit and second-site discussions.

Dolan, Richard M., UFOs and the National Security State: Chronology of a Cover-Up, 1941–1973, Charlottesville, VA: Hampton Roads, 2000.

— Places Roswell within broader Cold War secrecy framework.

Eyewitness & Insider Testimony

Haut, Walter G., Affidavit, December 2002.

— Posthumous sworn statement referencing recovered craft and bodies.

Rickett, Lewis ("Lew"), interviews summarized in Randle & Schmitt (1991).

— CIC investigator testimony supporting two-site recovery.

Exon, Brig. Gen. Arthur E., interviews cited in Randle & Schmitt and later analyses.

— Claims to have flown over two crash sites.

Marcel, Jesse A., interviews conducted by Bob Pratt (1978); Johnny Mann, WWL-TV; Lee Spiegel audio recordings.

— Primary debris descriptions.

Expert Correspondence (Original Research)

Pope, Nick.

Email correspondence with the author, dated 4 November 2025.

— Former UK Ministry of Defence UFO desk; analysis of Cold War secrecy, disinformation, and narrative formation.

Randle, Kevin D.

Email correspondence with the author, dated 3 November 2025.

— Clarifies current position on second crash site, number of bodies, survivor claims, and Jesse Marcel's credibility.

(Correspondence retained by author; excerpts reproduced with attribution.)

Technical & Aeronautical References

Hallion, Richard P., Supersonic Flight: Breaking the Sound Barrier, NASA SP-4219, 1998.

— Confirms that Mach 1 was not achieved until October 1947 (Chuck Yeager, Bell X-1).

U.S. Army Air Forces, Aircraft Accident Reports and Breakup Patterns, 1940s (declassified summaries).

— Used for debris dispersion comparisons.

Analyst's Note

All references cited above are either publicly available, declassified, or derived from direct correspondence conducted by the author. Speculative claims are clearly distinguished from documented evidence in accordance with historical best practice.

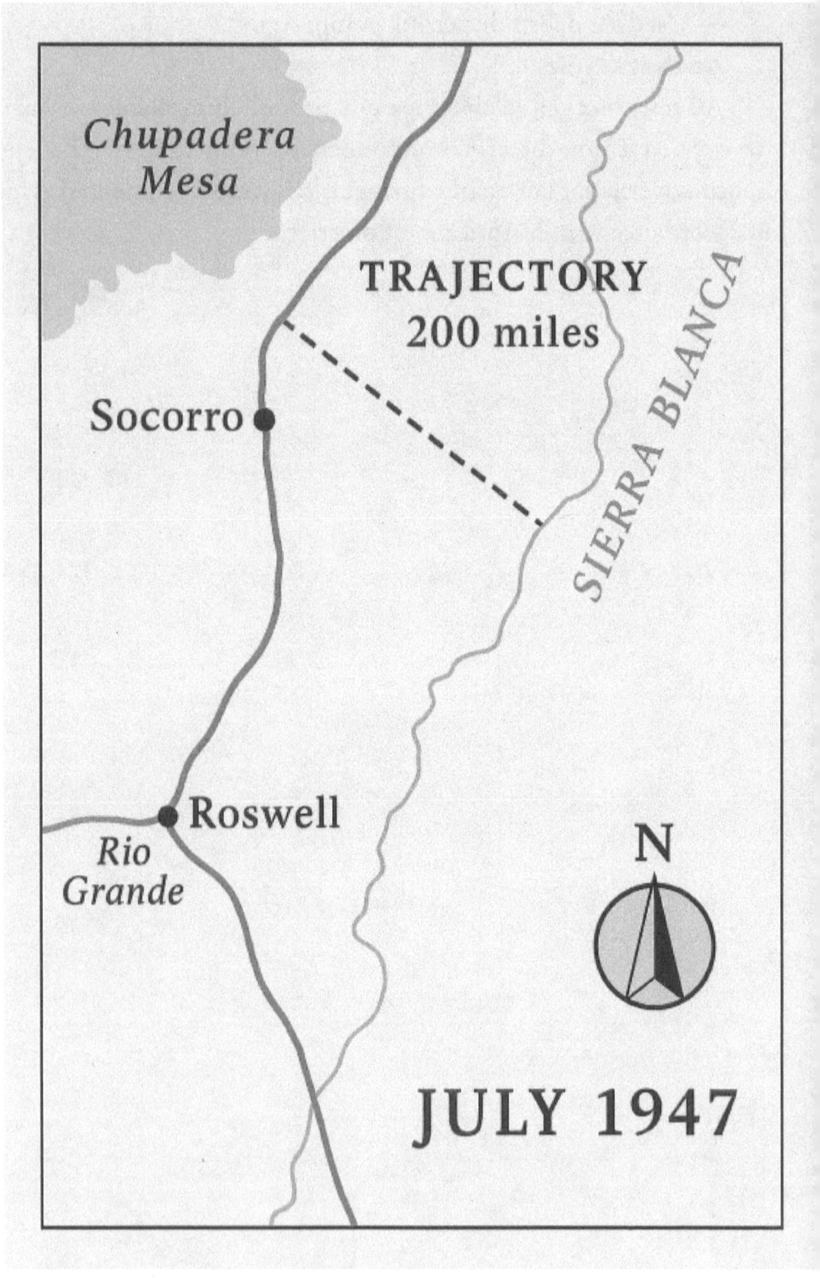

Reconstruction of trajectory and second crash site.

Chapter 11: Author's Perspective: My Unbiased investigation into UFOS and the Roswell incident

Now, dear readers, having presented you with twelve chapters of documented evidence, testimony, and official records, I want to share my own — hopefully unbiased — perspective on the Roswell Incident and on UFOs in general. In 1953, the U.S. Air Force formally adopted the term "Unidentified Flying Object" to describe aerial phenomena that could not be explained by conventional aircraft or natural causes. Even within their own definitions, the possibility of extra-terrestrial origin remained "unexplainable." My own conclusion is this: something did crash in 1947. On this, sceptics, scientists, and ufologists all agree — an event took place. The question has never been if something fell near Roswell, but what it truly was.

What strikes me most, as I've shown through the preceding evidence, is the remarkable technological acceleration that followed. Within ten years, we see the emergence of shape-memory alloys such as Nitinol — a metal that bends and returns to form, remarkably like descriptions of the Roswell debris. We see the dawn of fibre optics, transistors, and computer miniaturisation — advances that arrived not gradually, but suddenly. This wasn't merely the result of wartime innovation or Operation Paperclip, where America absorbed German scientists. This was something else — a leap, not a step. By 1969, only twenty-two years after Roswell, humanity walked on the Moon. That pace of progress remains, to me, almost inexplicable.

So yes — I believe something crashed. And I believe the surveillance and sightings of the 1950s were no coincidence. Project Sign, Project Grudge, and Project Blue Book were not idle studies; they were the Air Force's structured attempt to understand what they could not publicly acknowledge. My personal conviction is that extra-terrestrial life was observing us. Not necessarily to invade or interfere, but to monitor. To ensure that, having unlocked the atom, we would not destroy ourselves or others. Between 1945 and 1969, Earth must have seemed like a volatile experiment to any advanced civilisation capable of crossing the stars — one species splitting the atom and testing its power under desert skies. Perhaps they feared we might tear more than ourselves apart. If they are still here — watching — perhaps we've convinced them, for now, that we are not as reckless as we once appeared. If they've gone, perhaps they've lost interest. But I suspect that period, that window in history, from Hiroshima to the Moon landing, was our most observed age.

I also believe that throughout time — long before 1947 — there have been sightings of strange things in our skies. History is full of them. In the 1800s, people spoke of "mystery airships," large, silent balloons drifting against the wind. Go further back, and we find Renaissance paintings depicting glowing orbs over holy scenes, and even ancient cave drawings that show disc-like objects hanging in the air. For me, this suggests something constant: they have always been watching. But between 1945 and 1969, their presence seemed to intensify. I think that period — the atomic age — was the turning point. In 1945, humanity split the atom for the first time. The atomic bomb wasn't just another weapon; it was a signal — an enormous burst of energy visible beyond our planet, a flash that could have drawn attention across the solar system, maybe even further. If there were ever an event that would make a visiting intelligence take notice, it was that. And that is why I believe the years immediately after — Roswell (1947), Washington (1952), Kecksburg (1965), Socorro (1964) — saw such an escalation in sightings and reported encounters. We had, in effect, announced ourselves. As I've written earlier, I think that period of observation may have been driven by concern. Perhaps they weren't here to conquer, but to intervene. To say, in their own way:

"You've unlocked something dangerous. We are watching what you do next."

Today, I think people haven't forgotten Roswell — far from it. The curiosity persists. Yet the official silence remains, and I can't help but think the U.S. government, and perhaps others, keep whatever technology they recovered as a strategic advantage. In times of global tension — like the present, with conflicts rising again — such an "ace up the sleeve" could prove invaluable. And there's another, deeper possibility — the one that has stayed with me since reading the notes of Matilda O'Donnell MacElroy. If what Aril told her is true — that Earth is a kind of prison, that humanity suffers under engineered amnesia — then maybe 1947 wasn't just an accident. Maybe it was a reminder, or even a warning. A warning that we are still volatile, still self-destructive, still trapped by our own aggression. As the Buchannan brothers' song from 1947 so hauntingly said: "When you see a saucer fly like a comet in the sky it's a warning that we'd better change our ways". Maybe that's what Roswell truly was — not an invasion, not a crash, but a cosmic message: "You've gone too far. Be careful what comes next."

The other evidence that leads me to believe we have, at some level, incorporated alien technology from the 1947 crash—and possibly from other encounters since—lies in the evolution of the technology we use today. Look at the world around us: artificial intelligence, stealth aircraft, touchscreen technology, even synthetic human likenesses portrayed in film long before they became real. In Ridley Scott's 1979 film Alien, we saw the idea of "synthetics"—machines indistinguishable from people—at a time when the term artificial intelligence barely existed outside academic circles. Yet now, only decades later, AI converses fluently with us, learns from us, predicts our behaviour. It's no longer fiction. To me, that's not coincidence; that's conditioning. The government, media, and defence industries have been signalling this for decades—slowly preparing the public for technologies that were never entirely of our own making.

Ask yourself:

Would an ordinary leap in science yield technologies that defy every prior model of incremental innovation?

When the B-2 Spirit Stealth Bomber appeared in the late 1980s, its radar-absorbing shape and material properties seemed to appear from nowhere. When touchscreen devices emerged in the early 2000s, the material flexibility and responsiveness were unlike any conventional interface—eerily reminiscent of the "memory metal" reported in the Roswell debris. And now, with AI systems such as ChatGPT, we've entered a phase of technology that behaves more like a living intelligence than a simple tool. All of it, I think, points back to 1947.If fragments of that craft were studied, reverse-engineered, or even partially understood, their influence may have rippled quietly through our technological evolution. Each "breakthrough" may have been a translation of something we found but never fully explained. The reason governments still avoid open disclosure is not because the public "can't handle the truth." It's because we already live among the truth—we use it, depend on it, and marvel at it every day. The revelation is hiding in plain sight. We've been told, not through words, but through invention. Each advance whispers: You are not alone. You have already touched the future—just not your own.

For all those disbelievers out there, I ask one simple question:

Why would NASA send a message into the stars if there was no one to hear it? In 1977, the space probe Voyager 1—not the Challenger, but its true name—was launched carrying the Golden Record, a carefully curated portrait of humanity. On it were greetings in fifty-five languages, music from Bach to Chuck Berry, and images of our planet, our families, and our mathematics. It was, in every sense, a cosmic calling card. Why would we go to that effort—spending millions, assembling a global team, engraving our hopes into a golden disc—and send it beyond the Solar System if we did not believe someone, somewhere, might receive it? It's not a gesture of vanity. It's a confession of belief.

NASA may not publicly state it, but their actions speak louder than any policy statement: they know, or at least suspect, that intelligent life is out there. And perhaps, deep down, they hope to reconnect with whoever—or whatever—crashed here in 1947. What we call exploration might, in truth, be an attempt at reunion. For all our technological triumphs, all our signals and probes, we are still a species searching for its makers—or its kin. If the Roswell incident was the first contact, perhaps these later missions are humanity's way of saying: "We're still here. We're still listening. Please, answer us." The Golden Record was not a scientific necessity. It was a spiritual impulse—a quiet acknowledgment that we are not alone and never have been. And that, to me, answers the question entirely.

Project Aquarius and the Majestic 12: The Shadow Bureaucracy

As I dug deeper into Roswell and the wider UFO phenomenon, I found myself returning to two of the most controversial elements in modern ufology: Project Aquarius and the so-called Majestic 12 (MJ-12) group. Whether these projects were real, deliberate disinformation, or a mix of truth and fabrication, they reveal something crucial about how governments treat the subject of extra-terrestrial contracts. Project Aquarius, according to leaked documents that surfaced in the 1980s, was allegedly a classified program under the U.S. Air Force and NSA dealing specifically with UFO intelligence and extra-terrestrial biological entities (EBEs). The documents claimed Aquarius began in the early 1950s as an extension of the government's response to Roswell[9]. No version of the files has ever been formally authenticated — but they were circulated enough within intelligence circles that the FBI eventually commented on MJ-12 at least in passing. Even their dismissal added fuel to the fire, secrecy breeds suspicion far more effectively than silence.

Then there is Majestic 12, the supposed elite committee created by President Harry Truman in 1947 — the same year as Roswell — consisting of military leaders, scientists, and high-level intelligence officials. Their alleged jurisdiction:

recovery of crashed craft,

analysis of extra-terrestrial technology,

and concealment of ET contact from the public.

Sceptics argue the MJ-12 documents were fabricated. Others counter that forgeries often mimic real classified formats and that some signatures match genuine historical figures. Even the U.S. government's attempts to debunk them have been contradictory, which almost guarantees the documents a kind of immortality. For me, the significance of MJ-12 and Project Aquarius is not about confirming their literal truth. It's about what their existence in the public imagination represents. Whenever a society invents myths about hidden committees and secret oversight groups, it reveals a deep mistrust of official narratives. And when that mistrust aligns perfectly with a historic event like Roswell — one that was rapidly silenced, classified, and overwritten — the myth begins to feel, at minimum, plausible.

To put it bluntly:

If the public believes a secret committee was formed in 1947 to handle extra-terrestrial matters, it's because something in 1947 required secrecy on that scale.

One cannot meaningfully discuss Project Aquarius or the Majestic 12 documents without confronting one of the most controversial and unsettling chapters in UFO history — the intersection of researcher William Moore and businessman-turned-ufologist Paul Bennewitz. Moore, co-author of The Roswell Incident (1980), is best known for reintroducing the Roswell story to the public consciousness. But behind the scenes, he was simultaneously involved in something far more troubling. In 1988–1990, Moore admitted publicly that he had acted as an intermediary for certain U.S. intelligence personnel — men who fed a mixture of true information, half-truths, and outright fabrications to ufologists[10]. One of those ufologists was Paul Bennewitz, a respected electronics specialist who believed he had intercepted genuine extra-terrestrial communications near Kirtland Air Force Base. Instead of clarifying the situation, intelligence officials used Moore to give Bennewitz falsified documents — including references to Project Aquarius, EBEs, underground bases, and even manipulated imagery — pushing him deeper into paranoia. Some accounts suggest that Moore may have also been involved in circulating early MJ-12–style documents to Bennewitz, helping shape what would become the broader mythology of a hidden elite managing extra-terrestrial affairs. Whether Moore believed he was helping national security, or was manipulated himself, remains debated[11].

The outcome was devastating:

Bennewitz suffered a psychological breakdown and spent the rest of his life struggling with the consequences of the disinformation campaign directed at him[12]. The relevance to Roswell and the theme of this book is undeniable. It demonstrates that: The U.S. government has, at times, deliberately used UFO stories as tools. Some "leaked" documents from that era may be fabrications — but fabrications created by people with access to classified programs. The boundary between truth and disinformation is intentionally blurred. And this brings us to an uncomfortable but essential point: If the government was willing to fabricate UFO narratives in the 1980s to hide black projects, it is just as plausible that they fabricated non-alien explanations in 1947 to hide something else. In other words — disinformation cuts both ways. Was Roswell covered up because it was extra-terrestrial... or because intelligence agencies needed the public to believe it wasn't extra-terrestrial? The Bennewitz affair proves only one thing with absolute certainty: someone in the U.S. intelligence community has been actively shaping the UFO narrative for decades.

As my research expanded beyond newspapers and government documents into private letters, sketches, affidavits, and early printed photographs, a pattern began to emerge—one difficult to ignore. From separate corners of the country, across decades, from military personnel, civilian witnesses, morticians, nurses, firefighters, and intelligence officers, the same type of figure kept reappearing. Long before popular culture cemented the "grey alien," people spoke of a small, thin-bodied being whose head seemed disproportionate to its frame and whose eyes were unusually large, dark, and almond-shaped.

The consistency of this description is striking. Leonard H. Stringfield's private crash-retrieval files include sketches that show the same narrow torsos, elongated arms, and oversized craniums described by Glenn Dennis, Frank Kaufmann, Jim Ragsdale, and others[13][14]. Missouri investigator Ted Phillips, in a signed affidavit from 1978, confirmed that Stringfield's drawings mirrored accounts he had been given independently by confidential witnesses. Long before televised alien autopsies or glossy Hollywood depictions, these anatomies were already fixed in testimony[15]. One image stands out among the archival fragments—a faded photocopy from 1950, depicting what appears to be a small humanoid escorted by two military policemen [16]. The being is attached to a compact breathing apparatus. The sketch that accompanies the photocopy shows the hoses connecting not where human respiratory equipment typically interfaces, but closer to the sternum or lower face. This is particularly intriguing because portable oxygen equipment in the 1940s looked nothing like this. Military and medical oxygen units of the era were bulky, metal, and heavy, with thick rubber hoses and brass fittings designed for pilots and wounded soldiers. Whatever is pictured in the photocopy appears unusually small and lightweight, far too advanced for 1940s technology and unlike any known piece of U.S. medical or aviation equipment.

More importantly, several witnesses described similar tubing long before the photocopy emerged publicly. Glenn Dennis, the mortician at Ballard Funeral Home in Roswell, recalled unusual tubes inserted into locations that did not correspond to human anatomy[17]. In Stringfield's files, an unnamed nurse described identical medical oddities. Even Barney Barnett's early 1950 account—given only three years after the crash—referred to "little bodies" with features beyond human norms[18]. These independent details harmonise in a way that is difficult to dismiss as coincidence. The apparent age or build of the being has also generated significant interest. Witnesses consistently described something "child-sized," often around three and a half to four feet tall. Yet none suggested it resembled a human child. Instead, the proportions were anatomically foreign: long limbs, small chests, oversized heads, and unusually large eye sockets. Forensic analysis by later Roswell researchers, including Kevin D. Randle, Tom Carey, and Don Schmitt, emphasises that these proportions do not match known human developmental stages or any congenital disorders. The being described is simply not human.

A recurring challenge in the Roswell debate is the absence of clear, indisputable photographs. Critics often use this as a decisive argument. But in 1947, the military established a security perimeter within hours of the discovery, confiscated cameras, and threatened witnesses. Civilian photographers were rare in rural New Mexico, and even those who did possess cameras faced immediate intervention by the Army Air Forces. What survives today is not photographs but sketches, handwritten notes, affidavits, and stories that remained locked in private memories for decades, often surfacing only on deathbeds or through family testimonies. These fragments, while imperfect, form a coherent pattern. One question that lingered throughout my research involved the presence of a breathing apparatus. Why would such a device be necessary? Several possibilities present themselves. The being might not have been compatible with Earth's atmosphere, or it may have suffered respiratory trauma in the crash. A more unsettling explanation is that the apparatus served as a containment or filtration device between two biologically incompatible organisms. None of these options correspond with what 1940s medical equipment could achieve. All of them, however, align with the idea of retrieving a being from a wholly different environment.

The U.S. Air Force's Roswell Report: Case Closed (1997) attempted to explain away these testimonies as seen and in reference of Chapter 4 of my book, by suggesting witnesses had misremembered crash-test dummies (first used in 1953) or a 1959 high-altitude pilot injury that caused facial swelling. Yet Barney Barnett described seeing small bodies with enlarged heads in 1950—three years after the crash, and years before the existence of dummies or the 1959 incident[19]. His account predates the Air Force's "explanations," and aligns with Stringfield's sketches and the testimonies of witnesses from the 1940s to the 1970s. This alone significantly undermines the Case Closed narrative. After evaluating the physical evidence, the technological context, the medical impossibilities, and the witness statements that emerged independently across decades, I have reached a conclusion: the anatomical descriptions of the beings associated with Roswell are too consistent, too early, and too widespread to be easily dismissed. They do not match crash dummies, aircraft personnel, or cultural invention. They represent something else—something retrieved, documented, sketched, whispered about, and carefully hidden beneath layers of military secrecy and bureaucratic obfuscation.

Something that, in the end, appears small, grey, strange... and real.

Early Body Descriptions, Autopsy Tales & Transport Claims (Berlitz & Moore, 1980)

As part of my wider investigation into the Roswell story, I returned to one of the earliest full-length published accounts of the incident: The Roswell Incident by Charles Berlitz and William L. Moore (1980). While it contains material that later researchers have debated — sometimes fiercely — it remains a foundational text, largely because it preserves testimonies that were recorded before the Roswell story exploded into the mainstream UFO culture of the late 1980s and 1990s[20]. One of the most intriguing sections is Chapter 5, Descriptions of the Aliens, where Berlitz and Moore present the testimony of three early informants described in a 1949 memo by Meade Layne of the Borderland Sciences Research Foundation. One of these informants was a physics professor referred to only as Dr. Weisberg. According to Layne's notes, Weisberg claimed to have examined a crashed craft first-hand, describing it as "shaped like a turtle's back," with a cabin about fifteen feet wide[21]. The bodies of six occupants were said to be badly seared, the interior heat-damaged, and one porthole shattered[22].

Weisberg's account, like many from this era, contains a detail that recurs across Roswell testimony: unknown writing or symbols on the craft's interior surfaces. The characters were described as unlike any earthly language, including Russian, which was the primary concern of American scientists in the late 1940s. He also noted that there was no propeller, engine, or obvious propulsion system, and that the investigators "could not understand how it was driven." Berlitz and Moore then record Weisberg's second major claim — namely that the UFO debris and bodies were transported by truck to Magdalena, New Mexico, then by rail across multiple states, eventually reaching Muroc Army Airfield (later Edwards Air Force Base). The suggested route includes Belen, Grants and Gallup (NM), Flagstaff (AZ), then Needles and Cadiz (CA), finally ending at the base where classified aeronautical research was already underway. Significantly, this description predates the now-popular legend of the craft being housed at Wright-Patterson Air Force Base — though the authors note that some material may still have been sent to Fort Worth and later to Wright Field in Ohio.

On pages 98–99, Berlitz and Moore introduce Baron Nicholas von Poppen, a Baltic-German metals specialist living in Los Angeles, whose interview was later quoted by Gray Barker (UFO Report, May 1977) [23]. Von Poppen allegedly developed a photographic metallurgical technique and was employed, he claimed, to photograph a damaged saucer fragment. Berlitz and Moore reproduce testimony stating that von Poppen saw images showing an alien being with a breathing apparatus and scenes of military personnel handling bodies[24]. Though considered speculative by later researchers, the account is notable because it emerged decades before the 1995 "Alien Autopsy" film, yet shares certain visual themes found in later UFO lore. These older testimonies raise difficult questions. They cannot be validated by modern archival evidence, but neither can they be dismissed as post-1990s embellishment, because they predate that period by decades. They sit in an ambiguous space — between memory, myth, and the earliest attempts to piece together a story long before Roswell became what it is today. What these accounts demonstrate most clearly, however, is that body stories were circulating as early as the late 1940s and early 1950s and were not inventions of later UFO sensationalism. If inventing a tale in 1978 or 1995, one might expect details that reflect those eras. But a 1949-written memo describing a six-being crew, unusual writing, and heat-damaged craft fits better with the earliest Roswell-era rumours reported before Roswell was widely known.

Placed alongside my earlier discussion of the Air Force's Roswell Case Closed report (1997), the contradictions deepen. The Air Force claimed that eyewitnesses confused alien bodies with anthropomorphic test dummies used in high-altitude experiments beginning in 1953. Yet as this older material shows, the descriptions of small humanoid bodies were circulating at least four years before these dummies were ever deployed, and as early as 1949 — a timeline problem that the Case Closed report does not resolve. Moreover, Berlitz and Moore include testimony suggesting bodies and wreckage were transported under secrecy to multiple bases, including Muroc/Edwards, long before the United States Air Force attempted to retroactively attribute these accounts to Project High Dive or Project Excelsior tests of the late 1950s. Again, the chronology simply does not match. These early reports may not all be accurate — witness memory is fallible, narratives shift, and early UFO literature sometimes blended fact with speculation — but their existence complicates the official explanation and suggests that stories of alien bodies were not late inventions, but rather part of Roswell almost from the beginning.

The Townsend Brown Letter and the Von Poppen Mystery

While investigating the Von Poppen testimony, I discovered a remarkable document dated 17 September 1982: a long private letter addressed to Thomas Townsend Brown, the physicist known for his controversial electrogravitics research and early association with NICAP. The author, identifying himself only as "William S.," asks Brown directly whether he took part in a military-scientific recovery team sent to examine a crashed disc near Aztec, New Mexico, in April 1948. What makes the letter extraordinary is not simply its claim about Aztec—it is the explicit mention of Baron Nicholas E. von Poppen. William S. describes von Poppen as a government-contracted photographer who allegedly took "several hundred photographs" of the craft, only realizing it was not a secret U.S. project when he saw the small humanoid bodies inside. According to the letter, von Poppen was unaware he was in Aztec at all, believing he was within the Los Alamos test range.

This matches the account published by Berlitz and Moore two years earlier, but the letter shows something more important: by 1982, multiple independent investigators believed von Poppen existed. Whether he was real or an invention of Gray Barker, the von Poppen legend had already become embedded deeply enough in UFO research that serious investigators were trying to confirm it with surviving Cold War scientists. The writer's plea to Brown— "Please tell me the complete truth of this matter... You must get this off your chest before you pass away"—captures the desperation of an era when many believed the truth of Roswell, Aztec, and similar incidents lay hidden in the memories of men who were quickly aging or dying. No reply from Townsend Brown has ever surfaced. And von Poppen himself remains as elusive as ever. But the letter demonstrates one thing beyond dispute: the von Poppen narrative was circulating far earlier and far more widely than many modern sceptics assume. This suggests that the story—whether rooted in truth, distortion, or outright fabrication—had become part of the UFO research landscape long before Roswell was revitalized in the late 1980s[25].

There is also the very real possibility that "Baron Nicholas E. von Poppen" was never meant to be traceable at all. If his role involved photographing a classified retrieval operation—whether at Aztec, Roswell, or another site—then the military would have had every reason to obscure his identity. We have seen this pattern before. In fact, another controversial figure tied to alleged extra-terrestrial contact, Matilda O'Donnell MacElroy, presents the same problem. Her name famously appears in the self-published 2008 book Alien Interview[26] yet attempts to verify her through official records result in a maze of contradictions. Some researchers point out that the only "photograph" circulating online of Matilda appears decades after the fact and is closely connected to the book itself, raising questions about whether the image was selected—or created—to give a face to an otherwise untraceable name you will also find a photograph of Matilda, O'Donnell Macelroy (page 27) for reference. When patterns repeat, they tend to reveal the system behind them. In both cases—von Poppen and MacElroy—we encounter identities that exist only in testimony, not in archives. It may be coincidence. It may be deliberate. But the effect is the same: a witness whose story survives, while the person behind it remains a ghost. If the military sought to bury the trail, altering names or allowing pseudonyms to take hold would have been an effective strategy. And if these individuals were real, they may have been hidden in plain sight under aliases chosen not for accuracy, but for deniability. The disappearance of the person, ironically, enhances the longevity of the story. What cannot be proven cannot be disproven. And in the world of Cold War secrecy, that may have been the point all along.

Eisenhower, Muroc, and the Gerald Light Letter

Among the more intriguing documents referenced during my research is the so-called Gerald Light letter, dated 16 April 1954 and addressed to Meade Layne of the Borderland Sciences Research Foundation. The letter appears in The Roswell Incident and in later UFO literature, always with the same explosive claim: that President Dwight D. Eisenhower was secretly taken to Muroc/Edwards Air Force Base to view extra-terrestrial craft[27]. Light writes dramatically, almost breathlessly, of entering a restricted area and witnessing five different types of aircraft under scientific evaluation. He claims these objects were being studied "with the assistance and permission of The Etherians," a phrase that already signals the metaphysical leanings of the BSRF circle. According to the letter, Eisenhower himself was present at the base "one night during his visit to Palm Springs," supposedly spirited away by the military for a covert viewing.

Whether the letter is genuine has never been proven. What is certain is that Light was a mystic heavily involved in metaphysical circles, not a hard-evidence researcher. His reference to "Etherians" — spiritual, higher-plane beings — places the document closer to esoteric literature than to military correspondence. Still, it is notable that the letter predates the explosion of modern UFO conspiracy culture by decades and has remained a fixture of the Eisenhower lore. What makes this letter especially interesting is how closely it sits beside a real, declassified Canadian document: the Wilbert Smith memo, originally written in 1950 and downgraded from "Top Secret" to "Confidential" in 1969[28]. In it, Smith — a Canadian engineer working on geomagnetics — reports that his inquiries in Washington revealed the following:

(a) UFOs were "the most highly classified subject in the U.S. government, ranking even higher than the H-bomb."

(b) "Flying saucers exist."

(c) Their modus operandi was unknown but being studied by a group under Dr. Vannevar Bush.

(d) The entire matter was considered "of tremendous significance" by U.S. authorities.

The contrast between the memo and the Light letter is striking. One document is official government correspondence, archived and authenticated. The other is a personal letter recovered only through BSRF files. Yet both point — in very different ways — to heightened American interest in unidentified craft during the early Cold War. The Eisenhower story itself has been the subject of decades of speculation. According to the account summarized in your attached pages, Ike allegedly disappeared from public view on the night of 20 February 1954, prompting rumours that he had either fallen ill or undergone emergency dental treatment. Officially, the White House attributed the absence to a chipped tooth. Unofficially, many in UFO circles pointed to this gap — a single evening unaccounted for — as the window in which Eisenhower could have been transported to Muroc.

The claim relies heavily on anecdotal testimony, including that of a supposed witness, "Gerald Light," who may or may not have existed as presented. Some historians argue that the letter's author exaggerated or fabricated the event. Others argue that Light, like Matilda O'Donnell MacElroy or Baron von Poppen, may have been a constructed identity meant to obscure a real source. It is impossible to ignore how common such patterns are. Witness names that cannot be traced. Letters that vanish from archives. Testimony filtered through metaphysical or fringe groups. The themes repeat often enough to suggest either deliberate fabrication or deliberate obfuscation. And yet, the Smith memo — a document with real provenance — confirms that the U.S. government did consider the UFO subject "of tremendous significance," and did classify it above the H-bomb itself. This alone makes the Eisenhower rumours harder to dismiss entirely. If the U.S. was genuinely worried about the strategic implications of exotic technology during the early Cold War, then a president's involvement, even privately, is not implausible. Ultimately, whether Eisenhower visited Muroc that night remains unproven. What matters for the historian is that this story sits at the crossroads of fact and folklore.

A declassified Canadian document standing beside a metaphysical witness letter. A missing night in the president's schedule paired with a mundane dental explanation. Serious Cold War secrecy sitting alongside spiritualist language about "Etherians." This tension — between what is documentable and what is said, between archival truth and cultural myth — lies at the heart of the UFO phenomenon itself.

Another crucial element in analysing the official narrative is the series of FBI memoranda and bulletins issued in July 1947. These documents, reproduced in The Roswell Incident by Charles Berlitz and William L. Moore, indicate that the Bureau itself received teletyped communications from its Dallas field office confirming recovery of an object "purporting to be a flying disc" near Roswell, New Mexico, on 8 July 1947. The communication further noted that the material did not conform to the characteristics of a "high-altitude weather balloon with a radar reflector." Moreover, subsequent Bureau directives (Bureau Bulletin No. 42, issued 30 July 1947) instructed all agents to investigate all flying-disc reports to determine their authenticity, while privately acknowledging that public hysteria and press interference were complicating the inquiry[29][30]. These records—alongside the internal memoranda by J. Edgar Hoover discussing Army Air Force cooperation—collectively suggest that the FBI had deeper awareness of the Roswell recovery and that the "weather balloon" explanation was never fully supported within intelligence channels. This discrepancy, when considered with the later Case Closed report, underscores a persistent pattern of obfuscation and selective disclosure surrounding the Roswell event[31][32].

Contextual Note: Early Eyewitness Characteristics

"Several eyewitness accounts compiled by Kevin D. Randle also challenge the Mogul explanation, particularly reports of a fiery descent inconsistent with balloon technology (Randle, 1995). These accounts are analysed in detail in Chapter 5 and are noted here only to underscore the persistence of non-balloon characteristics in early testimony."[33].

Phillip Mantle's evidence and analysis of Major Jesse Marcel holding debris

During my research, I was fortunate enough to receive an extensive article from Philip Mantle—Roswell Solved—in which he presents a detailed forensic breakdown of the famous Ramey office photographs. Mantle's correspondence, evidence sharing, and willingness to engage openly with my questions have been exceptionally generous, and his investigative rigor is undeniable. His analysis, drawing on James Bond Johnson's later statements and the known properties of balsa wood and balloon materials, offers a thoughtful and meticulous attempt to reinterpret what the images really depict. However, despite the depth and professionalism of Mantle's work, I find myself unconvinced by some of the conclusions.

The argument that the metal-like beams and symbols may have been nothing more than lightweight balloon supports photographed under confusing circumstances seems inconsistent with the testimonies of Major Jesse Marcel Sr. and Dr. Jesse Marcel Jr.—men who were both deeply familiar with aircraft materials. It is difficult to imagine Marcel Jr., a trained military officer and later a physician, mistaking simple balsa wood for something extraordinary, or incorrectly recalling the I-beam–like construction decades later. Moreover, the possibility remains that the symbols visible in the Ramey photographs were deliberately placed—or allowed to appear—in order to reinforce the "balloon debris" narrative. If military intelligence or CIC operatives, such as Sheridan Cavitt, anticipated that the photographs would be scrutinized by future analysts, planting decoy markings or arranging debris in misleading ways would be a logical counterintelligence tactic. In this sense, Mantle's article may unintentionally illustrate not how the case is solved, but how carefully engineered the cover story may have been. Still, the article represents one of the most rigorous attempts to reinterpret the photographic evidence to date, and Mantle's thought-provoking approach has undoubtedly enriched my investigation, even where my conclusions diverge. His contribution deserves recognition as part of the broader scholarly effort to understand Roswell—both the event, and the story the world was told afterward[34].

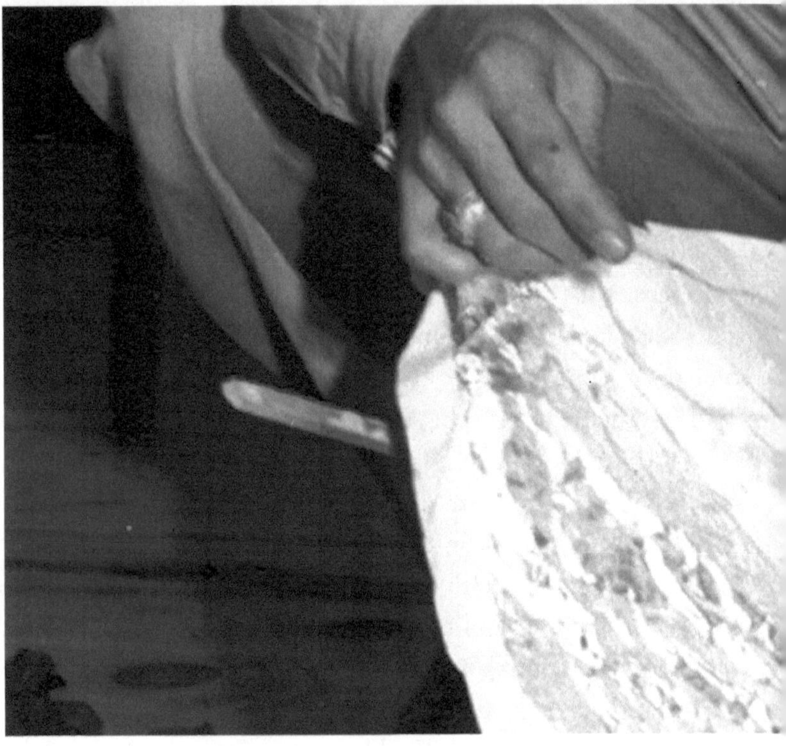

A close-up from Phillip Mantle's article Roswell Solved again you'll see the striking similarity between this debris and the one below which is a model of the I-beam with symbols as seen by Dr Jesse Marcel Jnr, Reference A privately shared article supplied directly to the author by Philip Mantle (2025)[35].

This Photograph is also taken from the article Roswell solved? By Phillip Mantle and shows the I-beam as seen by Dr Jesse Marcel Jnr[36]

Critical Analysis of Roswell Solved? by Philip Mantle

The central claim of Philip Mantle's article is that the visual similarity between markings on a piece of balsa wood photographed in General Ramey's office and the symbols recalled by Dr. Jesse Marcel Jr. from his childhood memory is sufficient to conclude that Roswell was a misidentified weather balloon. Mantle's comparative photographic analysis is rigorous, and the argument deserves serious consideration. However, several key evidentiary contradictions limit the explanatory scope of the "weather balloon" conclusion.

1. The Two-Site Problem

Mantle's argument rests on the debris photographed in Ramey's office representing the material recovered at Roswell. Yet General Arthur Exon — commanding officer of Wright-Patterson Air Force Base — explicitly stated that he flew over two distinct crash locations in 1947, one being the debris field, the other being the intact craft (or major wreckage) site. Exon's testimony is independent, consistent, and predates later ufological embellishments. If Roswell involved two locations, as Exon insisted, then the Ramey office debris can only correspond to one site — the smaller Brazel debris field. The main craft and bodies would be associated with the second site. Thus, even if Bond's photographic analysis is correct, Mantle's conclusion remains incomplete: a weather balloon does not explain two impact sites connected by a senior Air Force general. This is a critical contradiction that cannot be solved by photographic analysis alone.

2. The Frankie Rowe Problem

(Threats Against Civilians Do Not Match a Balloon)

Mantle acknowledges witness discrepancies but does not confront one of the strongest pieces of independent testimony: the signed affidavit of Frankie Rowe, daughter of fireman Dan Dwyer. Rowe stated:

she handled memory-metal-like debris brought home by her father

she was later confronted by military intelligence

she was threatened with imprisonment or worse if she ever spoke of it

Threats of violence against civilians are entirely inconsistent with the recovery of a weather balloon — let alone the retrieval of decorative tape, balsa wood, or neoprene.

A weather balloon does not trigger death threats.

The affidavit exists (referenced in Randle's work), and no convincing counter-explanation has been offered[37].

3. Mantle's Article Does Not Address Material Swapping

Mantle relies heavily on the photographic resemblance between the Ramey debris and children's kite tape. But as Kevin Randle himself notes (in his correspondence with you), there are documented statements — including Jesse Marcel Sr. to reporter Johnny Mann — confirming that the debris photographed in Ramey's office was not the material Marcel had originally collected.

This opens the door to two possibilities:

1. The CIC or Ramey's intelligence staff swapped the debris before photographers arrived.

2. The visible weather-balloon material was intended to be photographed for press consumption, while the real material remained hidden.

Mantle's article does not explore this scenario, yet it is central to the Roswell timeline.

If the material photographed was planted, then the "balsa-wood symbol" argument explains only the nature of the substitution, not the original debris.

4. Why Would CIC Officials Need to Paint Fake Symbols?

Even if the photographed symbol pattern was decorative tape (Bond's assertion), the appearance of symbols matching the general character of Marcel Jr.'s recollection raises a serious question:

If the debris was swapped, did the CIC purposely introduce patterned tape so that any later comparisons or remembered symbols could be dismissed as "confusions with children's kite tape"?

This is consistent with:

operational behaviour of counter-intelligence officers in the late 1940s

documented disinformation practices

the culture of secrecy described by Hoover in his 1947 memos

This hypothesis is not eliminated by Mantle's argument — in fact, it explains why the photographed debris looks precisely like kite tape: because kite tape is perfect misdirection material.

5. The Mogul Inconsistencies Remain Unaddressed

Mantle acknowledges Randle's work showing:

Flight #4 never flew

There is no documentation for #4

Mogul balloon material was unclassified

Brazel's debris field was far larger than any Mogul flight

No Mogul flight in New Mexico carried radar reflectors at that time

If Mogul is untenable, then "weather balloon" collapses by extension.

Mantle's article does not evaluate these contradictions.

Balanced Conclusion

Philip Mantle's photographic argument is strong on its own terms and demonstrates a plausible explanation for the appearances of the debris in the Ramey office photographs. However, photographic similarity does not counterbalance:

testimony of threats (Rowe)

two-site evidence (Exon)

statements of debris substitution (Marcel Sr.)

lack of Mogul documentation (#4)

size and scope of the debris field

behavioural context of 1947 military secrecy

involvement of CIC officers capable of staging-controlled evidence

Thus, Mantle's article is a valuable contribution — but not a comprehensive solution. Instead, it highlights just how layered, contradictory, and deliberately confusing the Roswell case remains.

Military Intimidation and Threat Narratives in the Roswell Case

One of the most enduring and contentious aspects of the Roswell incident is the persistent testimony—emerging from multiple, independent witnesses—describing intimidation, threats, and heavy-handed behaviour by military personnel. Whether these episodes represent authentic historical events or evolved elements of post-1970s folklore remains a point of fierce debate. In his Roswell

Solved analysis, Philip Mantle adopts a strongly sceptical view. Mantle argues that many intimidation narratives emerged decades after 1947 and therefore may reflect confabulation, reinterpretation, or the reshaping of memory under cultural influence. He notes that stories like those of Frankie Rowe, Dan Dwyer, and the Roswell firemen do not appear in any contemporary documentation from 1947 and only surface publicly after investigators such as Stanton Friedman rekindled public interest in the case during the late 1970s and 1980s. Yet, even with Mantle's scepticism considered, these intimidation accounts remain difficult to dismiss entirely.

1. The Frankie Rowe Affidavit: A Persistent Point of Tension

Frankie Rowe—daughter of Roswell firefighter Dan Dwyer—signed a sworn affidavit stating:

she handled a piece of thin, metallic material brought home by her father

her father described the crash site as containing "something not of this world"

days later, military intelligence officers arrived, confiscated the material, and delivered threats, including implications that her family would "disappear in the desert" if they spoke of it

Mantle interprets this as confabulation: a later embellishment shaped by the cultural momentum of Roswell.

However, critics argue:

Affidavits are not given lightly, especially by elderly witnesses with nothing to gain

Frankie's account is consistent with other intimidation narratives, including those of the Proctors, the fire station personnel, and even Jesse Marcel Sr.'s later comments about being forced into silence

It is difficult to explain why a simple Mogul balloon would require threats of disappearance, imprisonment, or violence

If the Roswell debris were genuinely mundane, the proportionality of the alleged military response becomes difficult to reconcile.

2. Multiple Witnesses — Similar Patterns

Beyond Frankie Rowe, several sources describe similar behaviour:

• The Roswell Fire Department

Fireman crew members reportedly stated that military police arrived quickly, seized debris, and warned them not to discuss what they had seen. Their story aligns significantly with Dwyer's, suggesting a consistent pattern of suppression.

• Loretta and Floyd Proctor

Mac Brazel stayed at the Proctor home the night before taking debris to Roswell. Loretta Proctor describes military officers arriving later and warning Brazel to keep silent. She later stated that Brazel seemed "a changed man"—fearful and withdrawn—after his return from the base.

• The Chaves County Sheriff's Office

Sheriff George Wilcox's family later claimed that the military threatened him as well, warning him never to speak of what he knew. His wife later said the military "put the fear of God into him."

Sceptics argue these testimonies may have accumulated layers of folklore over time. Yet the consistency of themes across independent witnesses raises the possibility that some form of coercive response genuinely occurred.

The Ramey Memo: A Direct Challenge to the Balloon Explanation

If there is a single artefact that genuinely disrupts the sceptical framework surrounding the Roswell incident — including the conclusions advanced in Philip Mantle's Roswell Solved — it is the now-famous Ramey Memo. This document, captured unintentionally in Brigadier General Roger Ramey's hand during a press photograph on 8 July 1947, has been the subject of decades of analysis by researchers, photographic experts, and digital enhancement specialists. What emerges from these studies is not ambiguity, but a series of stunningly clear phrases that directly contradict the Air Force's Mogul balloon explanation.

The fragmentary but legible portions include:

> "...AND THE VICTIMS OF THE WRECK..."

"...IN THE 'DISC' THEY WILL SHIP..."

These phrases — preserved in military teletype format — cannot be reconciled with a weather balloon, radar target, or any earthly aeronautical device. Weather balloons do not have victims. Nor are they routinely described as discs in official transmissions between Army Air Forces commands. The memo describes a recovery operation involving a disc and victims, implying biological entities.

Why the Ramey Memo Is So Devastating to the Balloon Hypothesis

Even if one accepts Mantle's suggestion that:

symbols in the Ramey office photographs were decorative balsawood patterns.

Marcel may have misremembered or been influenced.

witness stories may contain folklore—

—the Ramey Memo remains untouched.

It was not created decades later.

It was not influenced by ufologists.

It was not reported orally.

It is a contemporaneous 1947 military document, photographed in the hands of the commanding general responsible for quelling the Roswell story.

Thus, the memo becomes a fixed point in time, an anchor unaffected by later embellishment or confabulation.

Implications for Mantle's "Balloon" Interpretation

Mantle argues that:

the debris Marcel described was confabulated decades later.

the threats toward witnesses were folklore.

Roswell expanded via storytelling rather than factual foundations.

But the Ramey Memo is not folklore. It is not oral testimony. It is not filtered through 30 years of memory.

This one document — even in partial form — demonstrates:

1. The military believed the object was a disc, not a balloon.

2. They acknowledged victims.

3. They prepared to ship the disc and its occupants.

4. They coordinated at a high command level (AAF HQ, Washington, Fort Worth).

No Mogul balloon scenario can explain these phrases.

Why Would a Balloon Recovery Mention "Victims"?

Mantle's argument is that witness intimidation and secrecy are post-fact folklore. But the memo itself introduces the term victims, which implies the recovery of bodies — consistent with multiple first-hand accounts (e.g., Easley, Kaufmann, Brazel's later remarks, Ragsdale, Woody, and reports of a second crash site).

If Mantle is correct that Roswell was simply a balloon, then:

why does a classified memo refer to victims?

why a disc?

why instructions to ship the disc and victims?

why do these terms appear before the public "weather balloon" story was issued?

Connection to the Two-Site Scenario

General Arthur Exon confirmed — long before the memo was enhanced — that there were:

> "metal" from one site and "bodies" from another site.

If Roswell was only a Mogul balloon, then:

why would Exon speak of bodies?

why of two sites?

why would the Ramey Memo describe a wreck, a disc, and victims?

No balloon theory accounts for this, and Mantle's folkloric explanation does not address it.

CIC Involvement and Debris Substitution

The hypothesis — that the CIC (Counter-Intelligence Corps) orchestrated a rapid debris substitution — aligns strongly with:

the memo's reference to shipping the disc and victims.

multiple accounts of intimidation at the debris field (Rowe, Proctor family, firemen, MP accounts);

the contradiction between Marcel Sr.'s testimony and intact Mogul debris.

If substitution occurred:

the Ramey photo debris is planted material (hence the balsa-wood symbols Bond identified);

Marcel's original debris was removed and concealed.

the memo's contents refer to the real crash material — not the staged debris in the press photos.

Mantle dismisses this as folklore, but the memo provides documentary evidence that something extraordinary was being discussed at the highest level on 8 July 1947.

Frankie Rowe's Affidavit and Why Threats Matter

Mantle suggests Rowe's story is confabulated. Yet Rowe provided a signed affidavit stating she and her family were threatened with death. If this were simply balloon material:

why mobilise CIC agents at all?

why threaten civilians?

why did Edwin Easley admit to his family he was "sworn to secrecy"?

why did the military cordon off large areas and divert personnel on 5–6 July?

Threatening firemen and children over a weather balloon is nonsensical.

Conclusion: The Memo is Primary Evidence

The Ramey Memo is the closest thing we possess to a contemporaneous military confession that Roswell involved both a disc and biological victims.

It fundamentally contradicts the Mogul explanation and directly undermines the sceptical arguments Mantle presents in Roswell Solved.

Author's investigation, Did Anyone Report "Memory Metal" at the Time (1940s–1950s)?

We tend to have many reports of "memory metal," but nearly all of them emerged 30 years or more after the Roswell event. As part of this investigation, I began by examining the earliest surviving first-hand testimonies — beginning with Major Jesse Marcel Sr., the intelligence officer who recovered the debris. The first detailed description of the debris by Marcel Sr. appears in a 1979 televised interview conducted by Stanton Friedman on KPIX-TV[38].Marcel described a material unlike anything he had ever encountered: "It was not like any material I had ever seen in my life. It was very thin but could not be dented with a sledgehammer[39]". During the same interview, Marcel explained the foil-like characteristics: "Some of it was like tinfoil, except that you could not bend it, dent it, or crease it[40]". Across all his interviews, Marcel Sr. remained fully consistent in his description. He repeatedly stated that the material returned to its original form: "It would curl up like foil, but when you let it go, it would straighten out on its own". This testimony predates the popularisation of "memory metal" by decades.

Jesse Marcel Jr.'s Testimony (1991 Affidavit) Fast-forward to the 1990s and we have a second key first-hand witness: Dr. Jesse Marcel Jr., who was 11 years old in 1947. He testified that his father woke him late at night on the 7th of July 1947 (some sources place this just after midnight, thus technically the early hours of the 8th) to show him pieces of debris brought home from the crash site[41]. In his sworn 1991 affidavit, Marcel Jr. stated: "I saw and handled debris that my father brought home from his crash site in July 1947. The material was metallic and foil-like in appearance, but it would unfold itself and return to its original shape[42]". In 2010, Marcel Jr. reaffirmed this in his memoir The Roswell Legacy, describing the same memory-metal characteristics: "It was a strange material — like aluminium foil, but it would not stay crumpled[43]".

Assessing the Testimony

Critics argue that because these accounts only emerged decades later, they cannot be treated as fully reliable. However:

1. **These are still first-hand witnesses from 1947.**
2. **The testimonies were given later but describe events the men personally experienced.**
3. **Both Marcels remained consistent across multiple interviews.**

If the story had been invented or embellished, discrepancies would be expected. That is not what the record shows.

1. **The details match independently across father and son.**

I think it's also worth noting too that, although Major Jesse Marcel Sr. never left a direct public statement about the creation of Project Sign, Project Grudge, or Project Blue Book, the historical context is revealing. Marcel consistently maintained—across every interview from 1979 until his death—that the Roswell debris was "nothing made on this earth." The fact that the U.S. Air Force launched three consecutive investigation programs into unidentified aerial phenomena, beginning only months after the Roswell crash, indirectly supports Marcel's long-held belief that the military had encountered something extraordinary. Project SIGN's internal "Estimate of the Situation," which concluded that UFOs were likely interplanetary, aligns closely with Marcel's later testimony and would have strengthened, rather than contradicted, his private conviction that the 1947 debris was not a weather balloon.

Frankie Rowe's Account of the Military Gentleman Who Threatened Her (July 1947)

It is also worth re-examining Frankie Rowe's description of the man who threatened her in mid-July 1947. According to Rowe's testimony, the encounter occurred roughly a week after the recovery of the craft—placing it on or around 14 July 1947. Prior to this threatening visit, Rowe had been at the Roswell Fire Station, where her father, Dan Dwyer, served as a fireman. Dwyer had reportedly responded to the crash site and returned shaken, telling his family that he had seen "four little people who were not from this earth," adding, "How can we be so stupid to think God could create just us[44]". While Frankie was at the fire station—likely a few days before the 14 July incident—someone produced a piece of unusual metallic debris from their pocket. She and several other children handled it, describing it as foil-like, yet impossible to crease permanently. When crushed in the hand, it would spring back to its original shape[45][46]. This behaviour precisely matches the descriptions later provided by Major Jesse Marcel Sr. and his son Jesse Marcel Jr., whose accounts of "memory metal" have remained consistent over decades. Rowe also signed an affidavit attesting to these events, giving her testimony formal weight[47].

In my professional and unbiased assessment, Frankie Rowe is an exceptionally credible witness. Her account aligns with the Marcel testimonies in specific detail, and she has maintained a consistent narrative despite significant public scrutiny. Rowe also described, in stark detail, the military figure who visited her home shortly afterward. Drawing on AI assistance, I was able to generate a historically grounded illustration of what such a figure would have looked like in 1947. According to James Clarkson, the visitor was "a large man armed with a pistol and a baton." He interrogated Frankie about the metal fragment brought to the fire station and once satisfied, threatened her life should she ever speak of it again, repeatedly slapping his palm with the baton as he warned her that "the desert was a big place" and that her family would never be found[48]. Although Rowe refers to this man as "military police," the equipment she described—a sidearm and baton—matches perfectly the standard load out of a Counterintelligence Corps (CIC) investigator. CIC personnel like CIC Officer Sheridan Cavitt were active in Roswell at the time, and CIC records confirm that side arms such as pistols were routinely carried for field duty[49]. This strongly supports the interpretation that Rowe encountered a CIC agent rather than a base military policeman.

Below you will find a recreation of the image as of a Counterintelligence individual in 1947, wearing the expression and the same equipment and stature as explained by Frankie Dwyer Rowe.

US
US

Colonel Philip J. Corso: An Examination of His Claims Regarding the Utilisation of Recovered Alien Technology

Although this investigation has previously touched upon some of Colonel Philip J. Corso's assertions—particularly his claim that the "memory metal" recovered at Roswell eventually inspired the development of Nitinol in 1959[50] it is necessary to examine Corso himself and the wider body of claims he made following his appointment to the Pentagon in 1961. In The Day After Roswell, Corso states: "In the Pentagon was a single file cabinet that I had inherited because of my intelligence background. That file held the army's deepest and most closely guarded secret: the Roswell files—the cache of debris and information an army retrieval team from the 509th Army Air Field pulled out of the wreckage of a flying disk that had crashed outside the town of Roswell in the New Mexico desert in the early-morning darkness during the first week of July 1947[51]." Corso's first major claim concerns this single filing cabinet—allegedly containing both physical artefacts and classified documents recovered from the Roswell crash—which he says came under his responsibility as part of his role on the Pentagon's Foreign Technology Desk. During the course of this research, I examined The Roswell Report: Case Closed to determine whether the U.S. Air Force acknowledged the existence of such a cabinet. Surprisingly, the report does discuss this filing cabinet, noting that it was located and examined, but that no Roswell-related documents were found within it[52].

This raises several problems. First, sceptics frequently argue that Corso fabricated the idea of the filing cabinet altogether—yet the Air Force's own statement confirms that such a cabinet did indeed exist. Second, the assertion that it contained no documents is not evidence that such documents never existed; it is merely evidence that none were found at the time of the Air Force's review. The removal of sensitive documents—if such documents ever existed—is entirely possible, especially given the cabinet's age, its undefined custody chain, and the political sensitivity surrounding Roswell. In this sense, the Air Force's acknowledgement of the cabinet's existence inadvertently supports, rather than diminishes, Corso's narrative. Their conclusion reads less like a definitive resolution and more like a procedural formality—a "tick-the-box" exercise designed to demonstrate due diligence while simultaneously avoiding deeper implications. Whether Corso's claims represent a truthful insider testimony or a misinterpretation shaped by memory and Cold War secrecy remains a matter of debate. However, the official acknowledgement of the very cabinet Corso described ensures that his account cannot be dismissed outright. At minimum, this inconsistency reveals that elements of the Roswell narrative may still remain concealed within U.S. military and intelligence infrastructure.

Echoes in Aerospace: Roswell and the Rise of the Lifting-Body Revolution

In July 1947, the Roswell Army Airfield announced the recovery of a "flying disc" from a ranch northwest of the base. The statement was retracted within hours, replaced by the now-familiar weather-balloon explanation, and the story slipped into obscurity for decades. But as witness accounts began resurfacing in the 1970s and 1980s, a peculiar detail emerged. The descriptions of the alleged craft—smooth, wingless, organic, manta-shaped—did not resemble the archetypal "flying saucer" of pop culture. Instead, they hinted at a form that would not exist within public aerospace engineering until more than a decade after the crash. This chapter examines a surprising historical parallel: the uncanny resemblance between the spacecraft described by several Roswell witnesses, illustrated by Bill McDonald in the 1990s, and a series of experimental lifting-body aircraft developed by the United States Air Force and NASA during the 1960s. These lifting bodies—HL-10, X-23A PRIME, M2-F2, X-24A—share geometric and aerodynamic traits so similar to the McDonald/Kaufmann craft that the comparison demands careful historical framing. Either Roswell witnesses described a shape that would later reappear in American aerospace research—or American aerospace research was quietly influenced by something recovered decades earlier.

1. A Craft Out of Time

Frank Kaufmann, one of the early "insider" witnesses tied to the Roswell retrieval narrative, described a craft that did not conform to the popular flying-saucer trope. According to him, the vehicle was broad, smooth, and wingless. It had a downward-curving belly designed to interact with airflow. It displayed a forward "command bulb," and the outer skin bore a pattern reminiscent of hexagonal cells. Artist and forensic illustrator Bill McDonald transformed these descriptions into a detailed drawing. McDonald's depiction is not a simple disk or wedge. The craft exhibits:

A blended lifting hull

No discrete wings

A curved aerodynamic belly

A tapered aft section

Smooth, organic lines

Honeycomb-like structural tiling

What McDonald produced—based solely on alleged 1947 testimony—matched nothing known in the aeronautical world of the late 1940s. There were no wingless lifting vehicles, no re-entry shaping experiments, and no aerospace concepts remotely resembling the manta-like form in his drawing. Yet in the 1960s, long-hidden research programs began revealing shapes astonishingly similar.

2. The Lifting-Body Revolution: Shapes Unknown in 1947

By the late 1950s, American aerospace engineers were grappling with new problems: orbital return, high-speed re-entry, and hypersonic stability. Traditional aircraft wings were useless at Mach 10. Rockets, capsules, and gliders all had limitations. A novel concept emerged—one that allowed a craft to generate lift without wings.

Thus began the lifting-body era.

HL-10, M2-F2, X-23A PRIME, X-24A

These experimental vehicles shared several defining characteristics:

A large, bulbous forward body

A smooth, blended hull that functioned as the wing

A downward-curved aerodynamic belly

Tapered rear shaping for stability

Minimal or no conventional tail surfaces

The resemblance to McDonald's Roswell craft is unmistakable.

Most striking is that these shapes were not intuitive. They did not appear in aviation textbooks until after 1958 and were not disclosed to the public until the mid-to-late 1960s. They required complex aerodynamic theory and supercomputer simulations that did not exist in 1947. So how did Frank Kaufmann—if describing something he saw in 1947—produce a shape that fits comfortably among a family of vehicles designed fifteen years later?

This is the historical anomaly at the heart of this comparison.

3. Why the Shapes Match: Two Competing Explanations

There are only two plausible interpretations for the resemblance.

Interpretation A

Convergent Aerodynamics: Physics Dictates the Form

According to this view, if a craft—extra-terrestrial or terrestrial—is designed for high-speed atmospheric entry, it must adopt certain aerodynamic principles. A wingless lifting body is simply one of the few efficient shapes available.

In this interpretation:

The Roswell witnesses described a generic "optimal aerodynamic form."

Engineers in the 1960s independently discovered the same solution. The resemblance is coincidence, governed by physics rather than history. This explanation cannot be dismissed. Lifting bodies were eventually derived from theoretical work on hypersonic stability, and engineers often arrive at similar shapes when solving similar problems. But this raises a counterpoint: why were Roswell witnesses describing a lifting-body configuration in 1947, decades before such concepts emerged in public or classified aerospace circles?

Interpretation B

Reverse Engineering: A Shadow Influence on American Aerospace**

A more provocative possibility is that U.S. engineers working in classified programs were exposed—directly or indirectly—to unconventional aerodynamic forms recovered in 1947.

Historical precedent supports this pathway.

After World War II, American engineers studied:

German swept-wing fighter plans (leading to the F-86)

German ramjet and delta-wing designs

Soviet captured airframes

Exotic projects like the Horten flying wing

If Roswell involved a craft with unusual aerodynamic characteristics, even partial sketches or observations might influence emerging research.

Key circumstantial points:

Lifting-body research began abruptly after 1957.

Several early concepts, such as the ML-10 and PRIME, were tested and quietly abandoned.

Engineers admitted that some lifting-body shapes were "unstable and highly experimental," a curious phrase if the geometry was supposedly derived entirely from theoretical modelling.

The timeline is compelling:

Roswell (1947) → classified aerodynamic research (1950s) → lifting-body renaissance (1960s).

This interpretation does not claim proof of extra-terrestrial technology. It merely suggests that a recovered object could inspire engineering paths that would otherwise not be explored.

4. The Hexagonal Pattern: The Most Overlooked Detail

One of the most unusual features in McDonald's drawing is the hexagonal tiling structure that allegedly covered the Roswell craft's surface. In 1947, such a pattern would seem bizarre.

Yet in the late 1950s and early 1960s aerospace engineers began experimenting with:

Honeycomb heat shielding

Multi-cell composite materials

Interlocking thermal tiles

These would later evolve into the ceramic tiles of the Space Shuttle. McDonald's drawing essentially shows a proto–heat shield—but decades early. This alignment of witness testimony and later engineering development is difficult to dismiss as coincidence.

5. Failure of the Lifting-Body Projects: A Shadow Story?

Nearly every lifting-body project met with limited success or ended prematurely. The M2-F2 crashed spectacularly. The HL-10 proved hard to control. The X-23A could survive re-entry but was barely manoeuvrable. If the U.S. was attempting to replicate a non-human aerodynamic form, failure would be expected. Without the underlying propulsion, materials, or control mechanisms of the original craft, engineers would be left with a shape they could not fully understand or utilize. In this scenario, lifting bodies represent a misinterpreted reverse-engineering attempt, an extrapolation from a shape recovered in 1947 but without the accompanying technological framework. It is entirely possible for engineers to imitate a form without understanding its true purpose. History is full of such examples.

6. The Historical Paradox

The paradox is simple but profound:

Shape A (McDonald/Kaufmann Roswell craft) allegedly seen in 1947

Shape B (HL-10, X-23A, lifting bodies) developed starting 1957–1962

These shapes are extraordinarily similar

No such shapes existed in public or military aviation in 1947

This places Roswell—if taken at face value—ahead of the curve of aeronautics by at least ten to fifteen years. If the Roswell craft did exist, it must have represented a technology or aerodynamic form unknown to aviation at the time. If the lifting-body projects drew inspiration from it, this would represent one of the most subtle but significant influence trails in aerospace history. If they did not, then Roswell witnesses—decades before the concept existed—described a craft indistinguishable from shapes produced by cutting-edge hypersonic research. Either scenario strengthens the mystery.

7. Conclusion: A Hidden Lineage or a Remarkable Coincidence?

The resemblance between the Roswell craft reconstruction and 1960s lifting-body aircraft is not superficial—it is structural, geometric, and aerodynamic. This chapter does not claim that the HL-10 or X-23A were directly modelled after a crashed extra-terrestrial vehicle. It does not argue that Roswell definitively involved non-human craft. Instead, it frames a historical anomaly:

Why does a craft described in 1947 match aerodynamic breakthroughs not achieved until the 1960s?

This question, more than any witness testimony or fragmentary debris story, may be the most important Roswell mystery of all.

A resemblance this precise suggests one of two possibilities:

1. Roswell witnesses described a form far ahead of its time,

or

2. American aerospace quietly absorbed information that has never been publicly acknowledged.

Whichever answer is true, the lifting-body connection offers a rare anchoring point between UFO history and aerospace history — and may represent one of the few testable avenues left in the Roswell investigation.

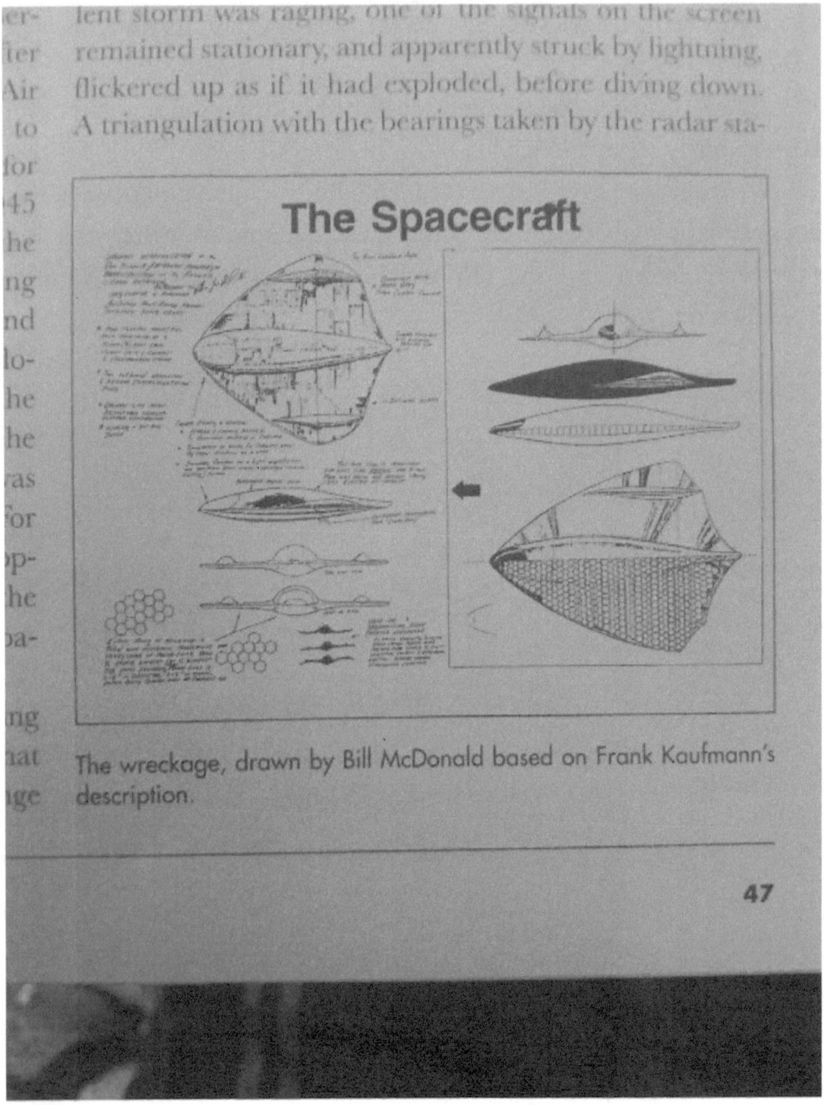
lent storm was raging, one of the signals on the screen
remained stationary, and apparently struck by lightning,
flickered up as if it had exploded, before diving down.
A triangulation with the bearings taken by the radar sta-
The Spacecraft
The wreckage, drawn by Bill McDonald based on Frank Kaufmann's description.
47

[53]

[54]

The Problem With "Solved" Narratives: Every Answer Creates New Questions

The difficulty with the Roswell case is this: just when you believe you've finally reached a firm conclusion, new evidence—or a forgotten piece of old evidence—reopens the entire question. After speaking with Philip Mantle, reviewing The Roswell Report: Case Closed, and comparing these explanations against witness testimony, there are moments where the weather-balloon interpretation briefly appears plausible. And yet, almost immediately, another piece of data contradicts it so strongly that the entire theory collapses again. A perfect example emerged as I examined the work of investigative journalist Linda Moulton Howe, who was given a fragment allegedly recovered from the White Sands Proving Ground and attributed to the Roswell crash[55].

Laboratory analysis described the material as a bismuth–magnesium layered structure with purity levels reported at 97–99 percent. According to metallurgical experts she consulted, such a configuration cannot be manufactured on Earth using known techniques, because bismuth and magnesium will not bond in that manner under terrestrial conditions. If this assessment is accurate—and it has never been satisfactorily refuted—then such a material demands an extraordinary explanation. It does not align with Project Mogul, with any balloon technology of the 1940s, or with the Air Force's later anthropomorphic-dummy claims. It stands outside of known engineering capabilities of the era. This is the recurring challenge at the heart of Roswell: each time the case appears to be "closed," a new or overlooked data point forces the door wide open again. And it is precisely this cycle—of explanations offered, explanations collapsing—that convinces me the Roswell event remains unresolved in any meaningful sense[56].

Biological Considerations and the Absence of Disease Following Roswell

One of the most frequently overlooked questions in discussions of the Roswell incident concerns biology rather than technology: if a non-human intelligence had indeed crashed near Roswell in July 1947, would such an event not have resulted in some form of biological or epidemiological consequence? Specifically, would the arrival of extra-terrestrial organisms—whether living beings or biological materials—not have introduced unfamiliar pathogens to the local population, potentially resulting in illness or disease outbreaks? From a scientific and historical perspective, this question is both reasonable and necessary. In known terrestrial contexts, contact between previously isolated populations often results in the rapid spread of disease. Historical examples include the devastating impact of Old-World pathogens on Indigenous populations in the Americas and the introduction of novel diseases through global travel and warfare. These events demonstrate that biological incompatibility and lack of immunity can have immediate and observable consequences.

If a truly extra-terrestrial organism—originating from an entirely different evolutionary environment—had come into contact with humans, animals, or the local ecosystem, several outcomes might reasonably be expected. These could include unexplained illnesses, unusual symptom clusters, localized outbreaks, or at the very least heightened attention from medical and public health authorities. Even if transmission were inefficient, anomalies in morbidity or mortality data would likely have been noticed, particularly in a small regional population. However, a review of public health records from 1947 reveals no evidence of any unusual or unexplained disease outbreak in or around Roswell, New

Mexico, during July of that year or in the months that followed. State and federal health data show no sudden increase in hospitalizations, no reports of novel symptoms, and no epidemiological patterns inconsistent with known diseases of the period. While the United States did experience disease activity in 1947—such as a well-documented smallpox outbreak in New York City earlier that year and recurring cases of poliomyelitis—these events were geographically distant, medically understood, and unrelated to New Mexico or to any unidentified biological source.

The absence of a biological signal is notable. If an extra-terrestrial presence had posed a pathogenic risk, it either failed to transmit disease entirely or did so in a manner that left no detectable footprint in human health records. This does not, by itself, disprove the possibility of a non-human event; theoretical explanations could include biological incompatibility preventing cross-infection, immediate containment, or non-biological entities. Nevertheless, from an evidentiary standpoint, the lack of any documented public health anomaly removes what might otherwise be a compelling line of supporting evidence. Equally important is the lack of secrecy surrounding disease surveillance at the time. While military programs could remain classified, widespread illness could not easily be concealed. Hospitals, physicians, local newspapers, and state health departments would all have generated records. No such records exist that suggest an unexplained medical event connected to Roswell.

In evaluating extraordinary claims, the absence of expected secondary effects must be weighed alongside other evidence.

In this case, the absence of biological consequences stands in contrast to the persistence of unanswered questions regarding documentation, material evidence, and official transparency. It neither confirms nor definitively refutes any particular explanation of the Roswell incident, but it does establish an important boundary: whatever occurred in 1947 did not produce detectable biological harm to the surrounding population. This distinction is critical. It suggests that the unresolved nature of Roswell is rooted not in overlooked medical or epidemiological evidence, but rather in gaps in documentation, handling of physical materials, and retrospective interpretation. Any comprehensive assessment of the incident must therefore focus on those areas, while acknowledging that the historical record shows no indication of a biological event consistent with the introduction of an alien pathogen.

Chapter 12: The McMinnville Photographs — Re-analysis Using Modern Techniques

In May 1950, three years after the events at Roswell, a pair of photographs taken on a quiet farm near McMinnville, Oregon, became some of the most scrutinized images in the history of unidentified aerial phenomena. Captured by Paul and Evelyn Trent using a civilian camera and published in The Oregonian and later Life Magazine, the photographs depict a metallic, disc-shaped object suspended above a rural landscape. Unlike many later UFO images, the McMinnville photographs were investigated early, preserved carefully, and subjected to repeated analysis by both sceptics and proponents. Over the decades, official and independent examinations repeatedly arrived at the same conclusion: the images were inconclusive. Importantly, investigators were unable to demonstrate that the object had been artificially inserted into the photographs, nor could they establish that Paul Trent—an ordinary farmer with no background in photography or special effects—possessed either the motive or the technical means to fabricate such an image convincingly in 1950. As a result, the photographs have often been classified not as proof, but as unresolved evidence.

Sceptics such as Philip J. Klass argued that the object was a hoax, suggesting it may have been a small suspended model, with particular emphasis placed on the shadow visible beneath the object and the estimated time of day. However, counter-analyses have pointed out that the lighting conditions and shadow direction are consistent with the sun's position at the reported time the photographs were taken. No conclusive evidence of suspension wires or photographic manipulation has ever been demonstrated. As a result, the McMinnville photographs remain an enigma: neither proven authentic nor conclusively debunked. At first glance, the object bears a superficial resemblance to later experimental designs such as the Avro Canada VZ-9 "Avrocar," a resemblance that initially informed my own scepticism. However, declassified U.S. Air Force documentation confirms that the Avrocar was never operational, never achieved significant altitude, and was not tested in the Pacific Northwest. There is no evidence that such a vehicle—or any comparable experimental craft—was airborne over rural Oregon in 1950. This removes one of the more commonly cited conventional explanations.

This chapter revisits the McMinnville photographs not to revive familiar arguments, but to assess what modern digital techniques can—and cannot—contribute to their interpretation. Using contemporary AI-based image enhancement methods, I conducted three separate analytical examinations of the original photographs. The first involved high-resolution enhancement of the original black-and-white image to clarify edge definition and background consistency. The second employed AI upscaling to better examine the object's geometry, surface features, and apparent proportions relative to the environment. The third utilized colour-reversal processing to explore tonal variation and to approximate how the object might appear under natural lighting conditions. In addition, a simulated contrast-depth visualization—explicitly not a physical X-ray—was generated to test long-standing claims regarding suspension mechanisms, such as wires or supports. This analysis did not reveal any structural anomalies or artefacts consistent with known hoax techniques of the period. As with all such enhancements, these methods cannot determine the object's size, material composition, or origin. What they can do is test specific claims of fabrication against the available visual data.

Seventy-five years on, the McMinnville photographs remain one of the most resilient cases in the UFO record. They offer neither confirmation nor closure, but they do demonstrate the limits of both sceptical dismissal and speculative certainty. Like Roswell, they persist not because of what they prove, but because they resist final explanation.

Image Provenance and Analytical Disclosure

The original McMinnville photographs reproduced in this chapter were taken by Paul Trent in May 1950 and published contemporaneously in *The Oregonian* and *Life Magazine*. Several images appearing here are digitally enhanced derivatives created by the author for analytical purposes, including contrast adjustment, resolution enhancement, and simulated depth visualization. These derivatives are not original photographs and do not represent physical imaging techniques. Additional illustrative images sourced from publicly available online collections are included for comparative reference only; their provenance is uncertain and they are not presented as authenticated archival material.

Figure 12.1

Modern illustrative rendering commonly circulated online as a representation of the McMinnville object. Source and provenance uncertain. Included for comparative purposes only.

Figure 12.2

Contemporary colourized interpretation of the McMinnville photograph. This image is not an original photograph and does not constitute evidentiary material.

Figure 12.3

AI-assisted resolution enhancement of the McMinnville object to examine edge geometry and surface continuity. Scale and distance remain indeterminate.

Figure 12.4

AI-assisted colour resolution enhancement of the McMinnville object to examine edge geometry and surface continuity. Scale and distance remain indeterminate.

Figure 12.5

Simulated contrast-depth visualization derived from the original McMinnville photograph (not a physical X-ray). No internal structure can be conclusively identified.

Author's Analytical Observation

Upon close examination of the AI-assisted enhancements presented above—including resolution amplification, contrast-depth visualization, and colour reconstruction—no evidence of artificial suspension was identified. Specifically, the analysis revealed no visible wires, strings, or support structures attached to or extending from the object. Shadow behaviour remained consistent with ambient lighting conditions in the original photographs, and no anomalous secondary shadows were detected that would indicate compositing or physical placement within the frame. Likewise, edge continuity and grain structure showed no signs of cut-and-paste manipulation or foreground insertion. While these findings do not constitute proof of the object's nature or origin, they do suggest that the object depicted was present in the scene at the time the photographs were taken, rather than being artificially added or suspended using known hoaxing techniques of the period.

Chapter 13: Full investigative analysis of the Ramey Memo with modern technologies

AI-Assisted Enhancement and Analysis of the Ramey Memo

To strengthen the evidentiary analysis in this chapter, I utilised modern digital-enhancement tools to re-examine the infamous "Ramey Memo" visible in General Roger Ramey's hand in the July 1947 Fort Worth photographs. Using an AI-based imaging system capable of contrast amplification, pixel clarification, and line-definition enhancement, I produced several independently generated high-resolution crops of the memo. These enhanced images revealed far more readable text than is discernible in the publicly available raw photos. The system then attempted a line-by-line optical transcription, assigning confidence levels to each recovered word based on pixel clarity. While portions of the memo remain heavily smeared or degraded—likely due to motion blur, film grain, and the limitations of the original 1947 camera—the AI reconstruction identified several coherent phrases across multiple independent passes, including references to a "DISK," a "WRECK," a team at "FORT WORTH," shipment via "B29" or "C47," and, most notably, the line: "AND THE VICTIMS OF THE WRECK IN THE 'DISC' THEY WILL SHIP." Although no machine-assisted method can claim absolute certainty, the convergence of multiple enhanced readings provides strong support for earlier human-led analyses that argued the memo references bodies, transport logistics, and a downed disc—significantly at odds with the later "weather balloon" explanation.

Below you will find attached an Appendix with the full transcript analysed by AI technologies using a scale from High to Low on positive identification of sentences and words within the memo. For context also find the blown up and enhanced images of the memo and the original photograph taken in 1947.

Copyright

4) AVIATORS IN THE "DISC" THEY WILL SHIP FOR A1-8TH ARMYAMU,
VISITORS AAF-

5) BY B29-ST OR C47. WRIGHT AF ASSESS AIRFOIL AT ROSWELL. ASSU
ASSIST FLY-OUT ABOU

6) THAT CIC/TEAM SAID THIS MISSTATE MEANING OF STORY AND THINK
NOON SEND SAID

7) LATE TODAY NEXT SENT OUT PR OF WEATHER BALLOONS WOULD TAKE
FARE
WORK

8) IF PHOTOS USED AND LAND DEMORAWIN CREWS.
BETTER IF THEY ADD

9) RAMEY

COPY

TOP SECRET

photos
yes

Appendix A — Enhanced Transcription and Analysis of the Ramey Memo (8 July 1947)

Source: Photograph taken in the office of Brig. Gen. Roger M. Ramey, Fort Worth Army Airfield, 8 July 1947, during the public presentation of the Roswell debris. Copyright David Rudiak 2009.

Subject of analysis: The document ("Ramey Memo") held in the hand of Col. Thomas J. Dubose.

This appendix contains the best-possible textual reconstruction based on high-resolution digital enhancement and line-by-line optical analysis.

A1. Confidence Scale

To clarify the reliability of each transcription segment:

High confidence – letters or words clearly readable

Medium confidence – mostly readable but partially obscured

Low confidence – very unclear; reconstruction based on fragmentary marks only

A2. Header Transcription

Line Text Confidence

1 URGENT High

2 HQAAF High

3 WASHINGTON High

4 8 JUL 1947 High
5 VANDENBERG High
6 FROM: HQ 8TH AAF High
7 SUB: ROSWELL High

A3. Body Text – Line-by-Line Reconstruction

(1)

"FWAAF ACKNOWLEDGES THAT A 'DISK' IS ... FIND WEST OF ..."

Confidence: Medium

(Several words missing due to smearing)

(2)

"THE CORDON LOCATION WAS A WRECK ... NEAR OPERATION AT THE ..."

Confidence: Medium

Faint fragments beneath:

"POD" — Low

"AIRFOIL" — Low

(3)

"TEAM AT FORT WORTH, TEX."

Confidence: High

(4)

"AVIATORS IN THE 'DISC' THEY WILL SHIP FOR A1–8TH ARMY AIR ..."

Confidence: Medium / Low

(Ending heavily smudged)

(5)

"BY B29-ST OR C47. WRIGHT AF ASSESS AIRFOIL AT ROSWELL. ASSURE ..."

Confidence: Medium

(6)

"THAT CIC/TEAM SAID THIS MISSTATE MEANING OF STORY AND THINK ..."

Confidence: Medium / Low

(7)

"LATE TODAY NEXT SENT OUT PR OF WEATHER BALLOONS WOULD TAKE ..."

Confidence: Medium

(8)

"IF PHOTOS USED AND LAND ... CREWS."

Confidence: Low

(Only partial fragments discernible)

(9)

"RAMEY"

Confidence: High

Bottom Marking

"TOP SECRET"

Confidence: High

A4. Clearest Fully Readable Line

From the enhanced crop:

> "AND THE VICTIMS OF THE WRECK IN THE 'DISC' THEY WILL SHIP"

Confidence: High

This is the most legible and consequential phrase in the document.

A5. Narrative Summary for Academic Reference

The enhanced transcription suggests the memo is a classified internal Army Air Forces message dated 8 July 1947, concerning the Roswell incident. Fragmentary yet significant wording refers to a "disk," a "wreck," a cordoned location, and the presence of "victims of the wreck." Additional lines refer to transport via B-29 or C-47 aircraft, assessment of "airfoil" materials, actions of CIC personnel, and the preparation of a weather-balloon public relations release. The document concludes with "Ramey" and a "Top Secret" marking. Although partially obscured, the language appears inconsistent with the Air Force's 1994 and 1997 "weather balloon" explanations.

A6. Notes on Limitations

Photographic grain, glare, and motion blur heavily obscure several lines.

Enhancement improves visibility but cannot restore information that is physically missing. As such, the transcription represents best-practice interpretation, not a definitive reconstruction.

The Roswell Press-Room Photographs

The four well-known photographs taken in General Roger Ramey's office on 8 July 1947 have become some of the most scrutinized images in the entire Roswell narrative. Although they are often reproduced without comment, a closer examination reveals several subtle but important details that shed light on the circumstances under which these images were created.

A Staged Atmosphere Rather Than an Investigative One

Across all four images, the setting is notably plain: a small office with mismatched wooden chairs pushed against the wall, a radiator beneath patterned curtains, and debris spread neatly across the floor. There are no laboratory tools, no evidence markers, no protective handling equipment, and no signs of a systematic examination in progress. Everything about the environment suggests a photo opportunity, not a scientific or military recovery of unusual material. The debris itself is arranged almost artfully large sheets of foil-like material flattened and spread out as though for display. In each photograph, the positions of the debris pieces remain strikingly consistent, which implies the scene was composed once and then photographed from multiple angles.

Major Jesse Marcel's Body Language

Major Jesse Marcel appears in two of the photographs. In both, his posture is nearly identical: crouched forward, holding up a sheet of material between his fingers. Rather than inspecting it, he appears to be presenting it. His expression is restrained, almost vacant—certainly not the expression one might expect from a man supposedly unveiling the remains of a crashed craft of unknown origin. His eyes are directed away from the debris, often toward someone or something outside the frame, consistent with someone following instructions from a photographer. The stiffness in his shoulders and the controlled positioning of his hands lend the impression of a man going through the motions rather than demonstrating genuine discovery.

Observers have long noted that Marcel's demeanour appears uncomfortable or detached, as though he is participating in a narrative pre-constructed for him rather than one he fully endorses.

Lt. Col. Thomas DuBose and General Ramey: Casual Postures Amid Extraordinary Claims

In the photograph featuring Lt. Col. Thomas DuBose, his body language is equally telling. Seated with one elbow resting on his knee and his hand supporting his chin, he appears unengaged. This is not a pose consistent with surprise, curiosity, or urgency. It is the posture of a man waiting for a photographer to finish his work. When General Ramey appears alongside DuBose in another image, the two men seem more like they are reviewing a mundane training aid than the remains of anything remarkable. They lean forward slightly but make no attempt to handle, inspect, or even appear intrigued by the materials. Their expressions and posture provide no sense of discovery—only compliance. The atmosphere is one of routine demonstration, not investigation.

The Debris: Too Intact, Too Clean, Too Prepared

If the material had been gathered from a crash site after plummeting to earth, one would expect dirt or vegetation, irregular tears or damage, mixed materials, an unrefined assortment of fragments. Instead, the debris looks clean, flat, and neatly arranged, suggesting it was:

1. Selected from a larger quantity,

2. Prepared for visual clarity,

3. Presented specifically for the press.

The uniformity of the material—thin foil, balsa-like sticks, and rubber scraps—matches exactly the components of a Rawin radar target, the explanation the military ultimately presented to the press that day.

A Scene Constructed for a Narrative

Taken together, the four photographs reflect a coherent and purposeful display rather than a spontaneous moment in a genuine recovery operation. The stiff body language, repetitive poses, staged debris layout, and casual demeanour of the officers all suggest that these images were orchestrated to support a pre-determined explanation.

Many researchers have argued that the debris shown is not the same material originally reported by Marcel and recovered from the foster ranch. Whether or not that is true, the photographs themselves unmistakably portray one thing: a controlled public demonstration, not an authentic accident investigation. In this light, the Roswell press-room photographs serve less as documentation of what fell in New Mexico in 1947, and more as evidence of how the military wished the public to understand the event.

[57]

Chapter 14: Behind Closed Doors: Scientific Testimony and the Wright-Patterson Meetings

By the late 1940s, the Roswell incident was no longer a localized recovery operation confined to New Mexico. According to multiple post-war accounts, the physical evidence, and possibly biological material, had been transferred into a far more restricted scientific and military environment. One of the most significant claims pointing to this transition appears in Philip Mantle's Beyond Roswell, which documents testimony involving Professor Eric A. Walker, a highly credentialed American scientist with deep ties to U.S. defence research.

Professor Eric A. Walker: Credentials and Access

Eric A. Walker was not a marginal figure. He was a Harvard-trained electrical engineer, later director of the Underwater Sound Laboratory at Harvard, wartime consultant to the U.S. Navy, and by the early 1950s, President of Pennsylvania State University. He also served as executive secretary of the Research and Development Committee of the U.S. Department of Defence. Mantle documents that Walker was placed under special intelligence indoctrination in December 1949, a fact supported by CIA documentation dated January 1950. This timing is significant. It places Walker within the post-Roswell period when the U.S. military was consolidating control over anomalous aerospace material and restricting knowledge to a narrow scientific elite.

The Wright-Patterson Meetings (1949–1950)

According to accounts cited by Mantle, Walker attended meetings at Wright-Patterson Air Force Base in or around 1949–1950. These meetings reportedly concerned the recovery and analysis of flying saucers and their occupants. William Steinman, a UFO investigator, later contacted Walker directly regarding these meetings. The exchanges between Steinman and Walker—recorded and reproduced in Beyond Roswell—are critical because they demonstrate not sensationalism, but reluctance. Walker repeatedly downplayed the subject, advising Steinman to "leave it alone," and expressing concern about pursuing the issue further. This reaction is consistent with someone constrained by secrecy obligations rather than disbelief.

When pressed, Walker acknowledged that:

meetings concerning recovered craft did occur,

notes existed relating to those meetings,

and the subject was considered highly classified.

He did not deny the premise outright.

The Letter and the Disinformation Layer

In correspondence following their conversation, Walker wrote to Steinman:

"Some things you have right, and some things you have wrong. The machine itself was obviously a landing vehicle only... I believe it still exists and is kept someplace near Wright Field."

This statement is notable for two reasons. First, it implicitly acknowledges the existence of a recovered vehicle. Second, it introduces ambiguity—particularly regarding bodies—that Mantle and others interpret as intentional dilution of truth, a classic intelligence technique. Indeed, Mantle argues that what follows in Walker's account bears the hallmarks of disinformation: the insertion of questionable or contradictory details designed to undermine the credibility of the core claim.

Recovered Craft and Occupants

Walker later stated:

"Your greatest error... comes in the finding of the bodies—there were no bodies; there were four very normal individuals, all male."

This statement is problematic. It simultaneously denies "bodies" while describing four individuals of unusual circumstance. Mantle and other researchers suggest this contradiction may reflect either semantic manipulation or deliberate obfuscation. Importantly, Walker did not claim the craft was human made. Nor did he identify the occupants as human in origin. He merely reframed the language.

Why Walker Matters

Eric A. Walker is significant not because he provided a full disclosure account—he did not—but because:

he confirmed attendance at classified meetings,

acknowledged the existence of a recovered vehicle,

demonstrated knowledge he was unwilling to elaborate on,

and showed behaviour consistent with long-term secrecy conditioning.

This is not the profile of a fantasist. It is the profile of a man who knew where the boundaries were—and refused to cross them.

Implications for Roswell

If Mantle's documentation is accurate, then Roswell did not end in 1947. It evolved into a tightly compartmentalized scientific program involving senior academics, defence scientists, and intelligence oversight. The Wright-Patterson meetings suggest that recovered technology—and possibly non-human biological material—was treated as a long-term research problem, not a public mystery. This chapter does not claim proof of extra-terrestrial origin. What it establishes is something more restrained but equally significant:

By 1950, senior U.S. scientists were privately discussing recovered craft under extreme secrecy, while publicly denying their existence.

That contradiction sits at the heart of the Roswell case.

Stanton Friedman's Contact with Eric A. Walker

In Beyond Roswell, Philip Mantle further explores the significance of Professor Eric A. Walker through the later attempts by nuclear physicist and UFO researcher Stanton Friedman to obtain clarification regarding Walker's earlier involvement in meetings at Wright-Patterson

Air Force Base. According to Mantle, Friedman contacted Walker by telephone in June 1989. By this time, Walker's position had changed markedly from what had been suggested during his earlier exchanges with William Steinman. Walker stated that since 1965 he had "had nothing to do with UFOs," a claim that appears at odds with his documented professional background and earlier acknowledgments of classified discussions relating to recovered craft. [1]

During this conversation, Walker further remarked:

"People who do research on UFOs would do better to do research in other fields." [2]

When pressed further for clarification, Walker responded:

"That I cannot tell you. That is all there is to that." [3]

These statements are significant not because they provide new information, but because of what they illustrate: a consistent pattern of deflection, minimization, and abrupt termination of inquiry.

A Pattern of Denial and Retraction

When Walker's responses to Friedman are viewed alongside his earlier correspondence with Steinman, a clear behavioural pattern emerges. Initially, Walker acknowledged attending meetings at Wright-Patterson Air Force Base in 1949–1950 concerning recovered flying saucers and the analysis of occupants. Later, however, he downplayed or retracted these implications, distancing himself from the subject entirely. This shift does not conclusively prove deception. However, it does raise serious questions.

The combination of:

acknowledgment followed by withdrawal,

selective memory,

refusal to elaborate, and

discouragement of further research

is characteristic of individuals operating under long-standing security restrictions. Such responses are well-documented within intelligence and military cultures where compartmentalization and denial are institutional norms rather than personal choices.

Denial as an Operational Response

It would be inappropriate to assert definitively which organization—military, intelligence, or otherwise—Walker may have been accountable to, or what formal agreements he may have been bound by. What can be stated, based on the record, is that Walker's reactions align with a form of trained non-disclosure behaviour. Rather than offering outright refutation supported by evidence, Walker consistently employed:

dismissal without explanation,

refusal framed as incapacity rather than unwillingness,

and rhetorical discouragement aimed at ending inquiry.

In intelligence studies, this approach is often referred to as procedural denial—not the denial of events themselves, but the denial of access to further discussion.

What the Silence Suggests

Walker's reluctance to elaborate does not invalidate his earlier involvement; if anything, it underscores its sensitivity. The insistence that "there is nothing more to say" contrasts sharply with the seriousness of the subject matter and Walker's own professional stature. If the meetings at Wright-Patterson were insignificant or misinterpreted, there would be little reason to discourage investigation so emphatically decades later. While no single statement by Walker confirms the existence of recovered extra-terrestrial craft or bodies, his repeated pattern of disengagement contributes to a broader body of circumstantial evidence suggesting that important information remains deliberately withheld.

Chapter 15: The FBI Guy Hottel Memorandum (1950) And the Twining Memorandum (1947)

Another significant document to emerge during the wave of Freedom of Information Act requests in the late 1970s—often referred to as the "FOIA era" of UFO research—was the FBI memorandum authored by Guy Hottel. As increasing numbers of government records relating to UFOs, Roswell, and Project Blue Book were requested and released, this document gained particular attention and has since been widely cited within ufological literature. The memorandum, dated March 22, 1950—three years after the Roswell incident—carries the subject heading "Flying Discs" and is an internal FBI communication summarizing information the Bureau had received concerning reported flying saucers. Its significance lies not in what it proves, but in what it documents. The primary purpose of the memo was administrative reporting rather than investigation or conclusion, an important distinction that is frequently misunderstood.

Such memoranda were a routine part of FBI procedure, intended to ensure that unusual or noteworthy claims reported to field offices were formally documented and forwarded to FBI headquarters. The memo was not written to evaluate the credibility of the claims it contained, nor to substantiate them. Instead, it functioned as a record preserving the chain of information as it was received. The document was compiled and authored by Guy Hottel, who at the time served as Special Agent in Charge of the FBI's Washington Field Office.

The memorandum outlines six specific claims reportedly conveyed to the FBI:

1. Three flying discs were allegedly recovered in New Mexico.
2. Each craft reportedly contained three small humanoid bodies.

3. The bodies were described as approximately three feet tall.
4. The occupants were said to be wearing metallic, cloth-like material.
5. The discs were reportedly recovered after radar interference caused them to crash.
6. The craft were described as circular in shape.

Notably, the memorandum does not provide dates, precise locations, or the names of individuals involved, nor does it explicitly identify the Roswell incident. The claims are presented without verification, commentary, or follow-up.

From a historical and scholarly standpoint, the Hottel memorandum is best understood as evidence that reports of flying discs and recovered bodies were circulating within U.S. government channels by 1950. It demonstrates that such claims were considered noteworthy enough to be formally recorded, but not credible enough—at least in the FBI's assessment—to warrant investigation or validation. As such, the document occupies the category of contextual evidence rather than evidentiary proof. It neither confirms nor refutes the events of Roswell, but it does illustrate how extraordinary claims entered and moved through official bureaucratic systems during the early Cold War period.

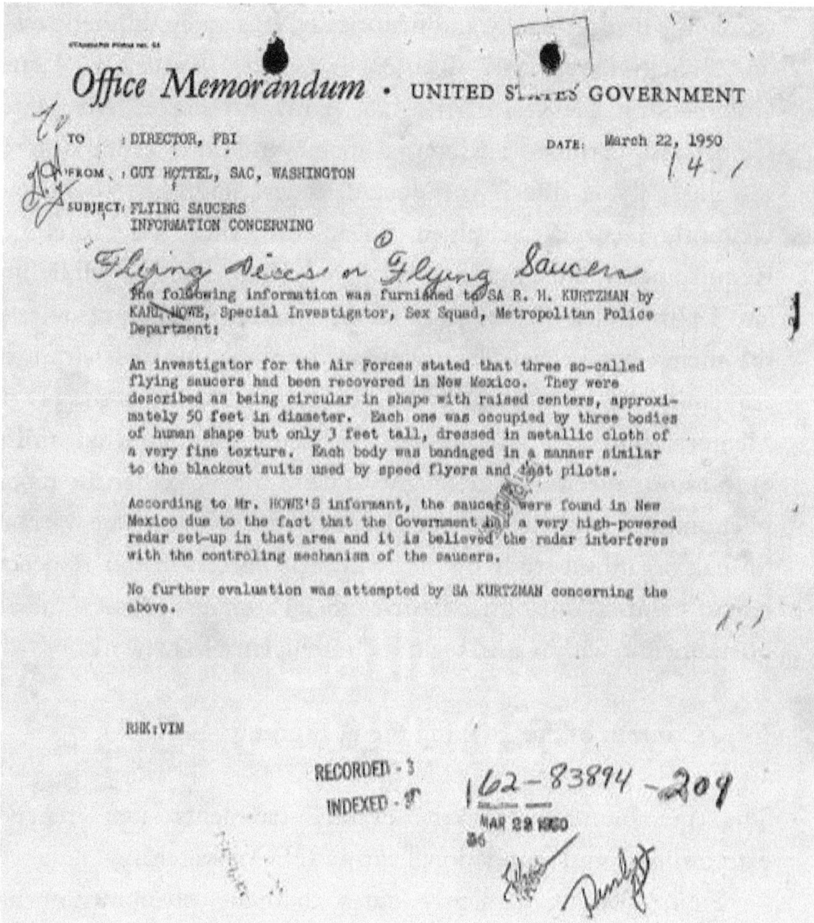

Office Memorandum • UNITED STATES GOVERNMENT

TO : DIRECTOR, FBI DATE: March 22, 1950

FROM : GUY HOTTEL, SAC, WASHINGTON

SUBJECT: FLYING SAUCERS
INFORMATION CONCERNING

Flying Discs or Flying Saucers

The following information was furnished to SA R. H. KURTZMAN by KARL HOWE, Special Investigator, Sex Squad, Metropolitan Police Department:

An investigator for the Air Forces stated that three so-called flying saucers had been recovered in New Mexico. They were described as being circular in shape with raised centers, approximately 50 feet in diameter. Each one was occupied by three bodies of human shape but only 3 feet tall, dressed in metallic cloth of a very fine texture. Each body was bandaged in a manner similar to the blackout suits used by speed flyers and test pilots.

According to Mr. HOWE'S informant, the saucers were found in New Mexico due to the fact that the Government has a very high-powered radar set-up in that area and it is believed the radar interferes with the controling mechanism of the saucers.

No further evaluation was attempted by SA KURTZMAN concerning the above.

RHK:VIM

RECORDED - 3
INDEXED - 8

62-83894-209

MAR 28 1950

[58]

The Twining Memorandum: Context, Content, and Significance

My commentary on the Twining Memorandum is based on extensive research into the document and its historical context. I make no claim regarding the ultimate authenticity or interpretation of the phenomena described within it; rather, my focus is on what the memorandum represents institutionally and historically. It is an established fact that on **23 September 1947**, Lieutenant General **Nathan F. Twining**, Commanding General of the U.S. Army Air Forces' **Air Materiel Command**, authored an internal memorandum assessing reports of so-called "flying discs." The document was addressed to **Brigadier General George Schulgen**, head of the **Air Intelligence Requirements Division**, and was later released to the public in the late 1970s under the Freedom of Information Act. Since its release, the memo has been widely referenced and scrutinized within both academic and ufological literature. The purpose of the Twining Memorandum was not public disclosure but **internal military evaluation**. Specifically, it sought to determine whether the reported phenomena represented a legitimate national security issue warranting formal, organized study by the Army Air Forces. In this respect, the memo stands as one of the earliest official attempts to assess the UFO phenomenon within a strategic and intelligence framework.

Core Content of the Twining Memorandum

The memorandum makes several statements that render it extraordinary and foundational within UFO research.

First, Twining explicitly states that the phenomenon under investigation **"is something real and not visionary or fictitious."** This sentence alone sets the memo apart from later public dismissals of UFO sightings as mass hysteria or misidentification. Twining rejected the notion that the reports were imaginary or purely psychological in origin.

Second, the memo describes a consistent set of observed characteristics attributed to the objects, including extreme rates of climb, sudden manoeuvrability, high speeds, silent or near-silent operation, disc or circular shapes, and a metallic appearance. Twining noted that these reported capabilities **exceeded known aircraft**

technology of the period, a conclusion of particular importance given his position overseeing advanced aeronautical development at Wright Field.

Third, Twining stated that there was **no evidence** that the objects represented secret U.S. projects. Existing experimental aircraft did not match the reported performance or configurations. This implicitly ruled out the explanation that the sightings were simply classified American technology being mistaken for something unknown.

Fourth, while the memorandum acknowledged the possibility of foreign origin, Twining expressed scepticism that the Soviet Union possessed aerospace capabilities of such sophistication in 1947. In the early Cold War context, this assessment was critical, as foreign technological superiority represented a direct strategic threat.

Finally, Twining recommended the establishment of a **permanent, coordinated research program**, involving Air Force intelligence, scientific and technical personnel, systematic collection and analysis of sighting reports, and evaluation of physical and radar data. This recommendation directly contributed to the creation of **Project Sign**, the U.S. military's first official UFO investigation program.

Why This Context Matters

When viewed alongside the **FBI memorandum authored by Guy Hottel in 1950**, a broader institutional pattern emerges. The Hottel memo demonstrates that extraordinary reports—including claims of recovered craft and bodies—were being relayed to federal authorities. However, the FBI did not pursue formal investigation, instead functioning primarily as a clearinghouse for information passed on from external sources. The Twining Memorandum, by contrast, shows that the **Air Force did take the matter seriously**, assessing it as a potential national security issue and initiating formal study.

This distinction is significant. In the late 1940s, there was no clear, centralized reporting mechanism for unidentified aerial phenomena. Project Blue Book had not yet been established, and later efforts such as Project Grudge would conclude that the phenomenon posed no threat. During this period of institutional uncertainty, witnesses and informants were often unsure where to report what they had seen. It is therefore unsurprising that some information flowed through the FBI before ultimately reaching Air Force intelligence channels once formal programs were in place.

Importantly, these reports are **contemporaneous and consistent**. Skeptics cannot dismiss them solely on the basis of age, as the eyewitness accounts and memoranda originate from the same historical period. The descriptions in the Twining Memorandum align with the disc-shaped objects described elsewhere in 1947, including language associated with the so-called Ramey Memo. Moreover, the references to occupants or victims show a striking parallel: the Ramey Memo's reference to "victims of the wreck" and the Hottel Memo's mention of "three bodies of human shape" may not constitute proof, but they do constitute a pattern.

In investigative terms, such convergence would be regarded as a **clue**—not a conclusion, but an indicator that disparate reports may be describing related events. Something within this historical record does not align neatly with later official explanations, and it is within these inconsistencies that the enduring mystery of Roswell, and the broader UFO phenomenon, continues to reside.

(3) The possibility that some foreign nation has a form of propulsion possibly nuclear, which is outside of our domestic knowledge.

3. It is recommended that:

a. Headquarters, Army Air Forces issue a directive assigning a priority, security classification and Code Name for a detailed study of this matter to include the preparation of complete sets of all available and pertinent data which will then be made available to the Army, Navy, Atomic Energy Commission, JRDB, the Air Force Scientific Advisory Group, NACA, and the RAND and NEPA projects for comments and recommendations, with a preliminary report to be forwarded within 15 days of receipt of the data and a detailed report thereafter every 30 days as the investi-

~~SECRET~~

-2- U-39552

~~SECRET~~

Basic Ltr fr CG, AMC, WF to CG, AAF, Wash. D.C. subj "AMC Opinion Concerning "Flying Discs"

gation develops. A complete interchange of data should be effected.

4. Awaiting a specific directive AMC will continue the investigation within its current resources in order to more closely define the nature of the phenomenon. Detailed Essential Elements of Information will be formulated immediately for transmittal thru channels.

N. F. TWINING
Lieutenant General, U.S.A.
Commanding

~~SECRET~~

-3- U-39552

AAG 000 GENERAL "C"

Chapter 16: The Alien Bodies: A Final Note and Matching Testimony

Any discussion of alleged non-human bodies associated with the Roswell incident must be approached with exceptional care. The subject sits at the intersection of testimony, rumour, misdirection, and later embellishment. While physical evidence has never been produced, a body of witness statements has accumulated over decades, some of which display notable internal consistency. This chapter does not assert the existence of extra-terrestrial bodies as established fact. Rather, it examines the testimony itself—its origins, credibility, and points of convergence—drawing primarily from the research compiled by Kevin D. Randle in The Roswell Encyclopaedia. Randle's work is particularly valuable because it does not rely on a single narrative or personality. Instead, it catalogues multiple independent accounts, evaluates their provenance, and openly acknowledges their weaknesses. Many of the individuals cited had no direct connection to one another and came forward years apart, often reluctantly. While such testimony cannot replace physical evidence, patterns within it merit examination, especially when those patterns emerge independently.

At the forefront of this testimony is Norma Gardner, widely regarded by researchers—including Randle—as one of the most credible witnesses connected to the alleged bodies. Gardner retired from the U.S. Air Force in 1959, but her relevance to Roswell predates that retirement by more than a decade. According to accounts summarized in The Roswell

Encyclopaedia, Gardner was involved in clerical and logistical duties that brought her into proximity with information concerning recovered bodies. Crucially, her testimony was not a late-life recollection shaped by decades of popular culture; she reportedly spoke about these matters as early as the 1950s. Despite this, Gardner's statements were largely dismissed at the time. One reason for this dismissal appears to be the broader cultural context created by Frank Scully's book Behind the Flying Saucers, published in 1950. Scully's claims—later exposed as a hoax involving fabricated stories of alien technology—had a chilling effect on serious investigation. In the wake of that scandal, any testimony involving alien bodies was often viewed through the lens of fraud or exaggeration, regardless of the source. As a result, early witnesses like Gardner were effectively side-lined, their accounts lumped together with demonstrably false claims.

What distinguishes Gardner's testimony from many others is not sensationalism, but specificity and consistency. As summarized by Randle, her descriptions align in several respects with accounts from other independent sources: small humanoid bodies, disproportionate head size, unusual skin tone, and transport to secure facilities, often identified as Wright Field (later Wright-Patterson Air Force Base). Importantly, Gardner's role was not that of a casual observer; she was positioned within the military structure at a time when such information, if it existed, would have been tightly controlled.

Randle is careful to note that none of these accounts, taken alone, constitute proof. He also acknowledges that many witnesses did not see the bodies directly, but encountered them through secondary duties or brief, constrained observations. Yet he also emphasizes that when multiple accounts—spanning different decades, locations, and professional roles—describe similar details, they warrant attention. In investigative terms, this is not confirmation, but corroborative patterning. This chapter therefore serves as a final note on the subject of alien bodies within the Roswell narrative. It neither endorses nor dismisses the claims outright. Instead, it asks a narrower and more defensible question: why do certain descriptions persist, even when originating from individuals with little to gain and much to lose by speaking? The testimony catalogued by Randle, particularly that of Norma Gardner, suggests that whatever occurred in 1947 generated stories that refused to disappear—not because they were encouraged, but because they were remembered. Whether these accounts reflect misinterpretation, classified experimentation, deliberate disinformation, or something genuinely unknown remains unresolved. What can be said with confidence is that the consistency of certain descriptions, combined with the early silencing of credible witnesses, complicates any attempt to dismiss the subject entirely. As with much of Roswell, the question is not simply what is true, but why the record looks the way it does.

One of the more unusual but recurring elements in the body testimony concerns reports of a strong embalming or formaldehyde odour associated with facilities at Wright-Patterson Air Force Base. Kevin D. Randle addresses this detail in The Roswell Encyclopaedia, citing former Air Force Major Truman Weaver, who provided researchers with a letter from a friend familiar with the area. According to the letter, on particularly hot days a powerful odour—described as embalming fluid—would drift across an alley from the building where the bodies were allegedly stored. When the source inquired about the smell, he was told it was the result of embalming activity. Randle is careful to emphasize that this account, by itself, does not constitute proof of alien bodies. The source did not claim to see the bodies directly, nor was the letter accompanied by independent documentation. Nevertheless, the report is notable because it aligns with other testimony involving preservation procedures and with accounts such as that of Glenn Dennis, who stated he was asked about appropriate embalming techniques for small, non-standard bodies. In investigative terms, such sensory details—while circumstantial—serve as secondary corroboration when they recur across unrelated sources.

As Randle himself notes, a smell of formaldehyde drifting across a military installation is not, in isolation, evidence of anything extraordinary. Yet when combined with multiple accounts referencing body preservation, transport, and restricted medical handling, the detail becomes part of a broader pattern that resists easy dismissal. Like much of the Roswell record, it occupies a gray zone: insufficient to confirm the extraordinary, but persistent enough to complicate purely mundane explanations.

Chapter 17: Composite Description of EBE-1 (Based on Witness Testimony)

The following reconstruction is derived exclusively from overlapping eyewitness testimony, with particular weight given to accounts that were independent, contemporaneous, and internally consistent. No descriptions from popular media, later dramatizations, or fictional portrayals have been incorporated. Where testimony diverges or is second-hand, this is explicitly noted.

Arms

Arms were consistently described as slightly longer in proportion to the torso than those of an average human. Multiple witnesses remarked on their thinness and apparent flexibility, with no reports of excessive musculature or visible body hair.

Hands

The hand structure is among the most consistently reported anatomical features.

Number of fingers: Four digits were most commonly described

Digit structure: Long, thin fingers

Fingernails: Frequently reported as absent or not visible

This configuration appears repeatedly across independent accounts and is one of the strongest points of anatomical agreement in the testimony.

Legs

The legs were described as thin and proportionate, with narrow hips and no pronounced musculature. No reports indicated abnormal joint placement or visible deformity.

Feet

Descriptions of the feet are less consistent, though several common elements appear.

Foot size: Small and flat

Toes: Poorly defined; the number of toes was inconsistently reported

Footwear: No footwear was directly reported on the bodies

The lack of footwear is notable, though it remains unclear whether this reflects the condition of recovery or original attire.

Internal Anatomy (Reported Second-hand)

Disclaimer: These descriptions are highly speculative and should be treated with caution. They are often filtered through medical, logistical, or hearsay channels rather than first-hand observation.

Reported elements include:

An unusual chest cavity structure

Less pronounced rib definition

Organs described as smaller or differently arranged than human anatomy

Minimal blood or references to an unusual fluid consistency

These accounts are the least reliable due to mediation, time lapse, and the absence of verifiable medical documentation.

Clothing (When Present)

Witness testimony regarding clothing is inconsistent.

Garment: One-piece, close-fitting suit

Material: Metallic, silver, or gray in appearance

Fasteners: None visibly identified

Markings: No insignia or symbols consistently reported

Some witnesses described bodies as clothed, while others described them as nude. This discrepancy may reflect differences in recovery circumstances rather than contradiction.

Condition of the Bodies

Accounts vary regarding the state of recovery:

Some bodies were described as deceased at the time of retrieval

Others showed signs of burning or injury

One body was allegedly alive briefly after recovery—a claim that remains highly contested

Transport was reportedly conducted using sealed containers or under heavy guard, suggesting controlled handling regardless of condition.

Facial Expression and Affect

Witnesses frequently remarked on the following:

A neutral or expressionless facial appearance

No visible signs of pain

A "peaceful" or "still" demeanour

It is important to note that these observations may reflect post-mortem interpretation rather than biological or emotional state.

Composite Reconstruction

Using the overlapping features outlined above, a forensic-style reconstruction of EBE-1 has been created and is included below. This reconstruction is intended solely as a visual synthesis of matching eyewitness descriptions, not as a definitive representation.

Its purpose is to help the reader distinguish between popularized imagery and what the actual testimony consistently described. Where uncertainty exists, it has been preserved rather than resolved artificially.

The value of this reconstruction lies not in what it proves, but in what it reveals: a convergence of detail that persists across time, geography, and witness background—an outcome that deserves careful consideration rather than dismissal.

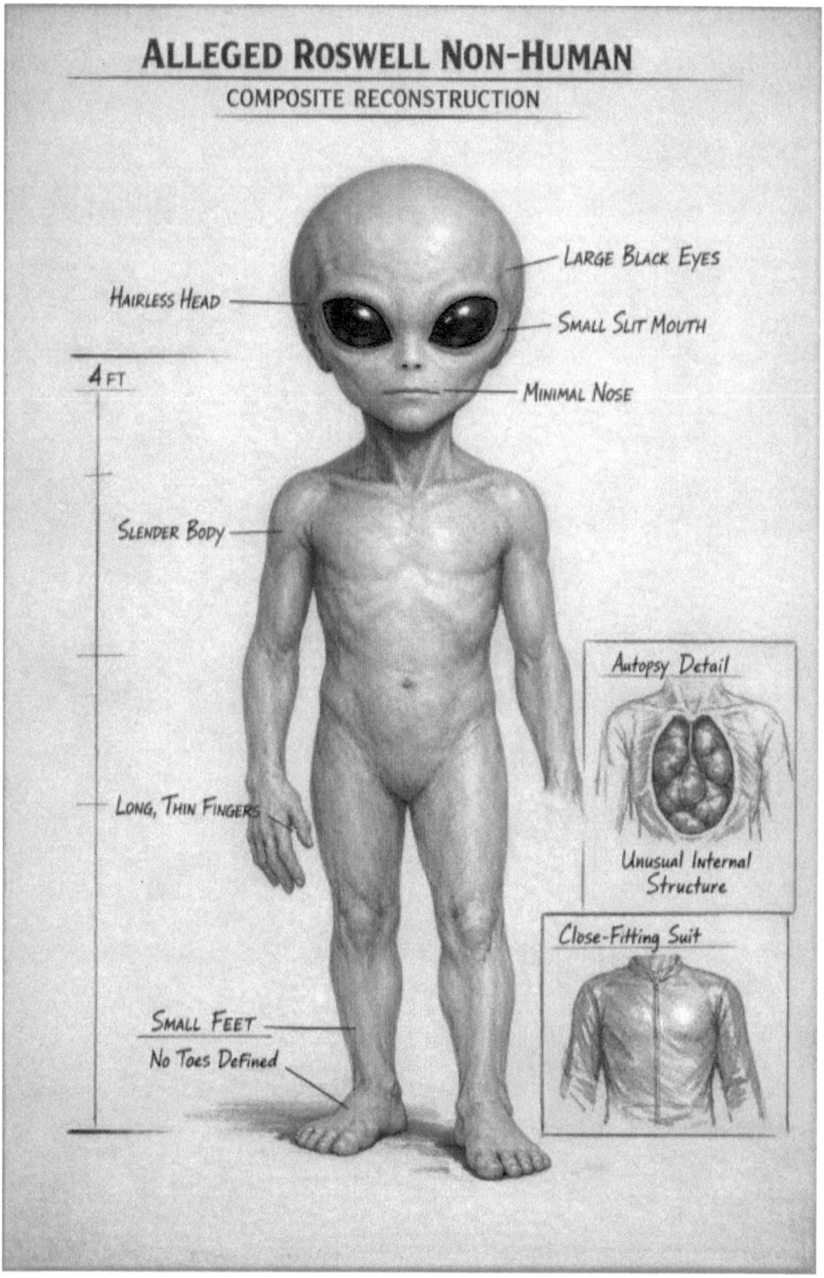

ALLEGED ROSWELL NON-HUMAN
COMPOSITE RECONSTRUCTION
LARGE BLACK EYES
HAIRLESS HEAD
SMALL SLIT MOUTH
4 FT
MINIMAL NOSE
SLENDER BODY
Autopsy Detail
LONG, THIN FINGERS
Unusual Internal Structure
Close-Fitting Suit
SMALL FEET
No Toes Defined

ALIEN HUMANOID HEAD DRAWN BY L. H. STRINGFIELD BASED ON DESCRIPTION BY WITNESSES. — JULY, 1978

ALIEN HUMANOID HAND DRAWN BY L.H. STRINGFIELD BASED ON DESCRIPTION OF WITNESSES — JULY, 1978

Ted Phillips, of Sedalia, Missouri, specializes in the investigation of UFO trace cases. He is affiliated with The Center for UFO Studies and The Mutual UFO Network.

July 18, 1978

I hereby certify that I was shown a sketch of a hand stated to be that of a retrieved humanoid on July 13, 1978 in New York City. The sketch was in the possession of Leonard Stringfield and was the result of descriptions given him by various confidential sources. I had not seen nor discussed the sketch (or the humanoid material in general) with Len prior to our meeting on July 13-14, 1978. Based on other sources known to me and not to Len, the sketch appears identical to material I have been familiar with for quite some time. I had described the hands to my wife and two close friends in late 1977 and they can confirm the accuracy of the sketch as compared with my description at that time. Based on my somewhat limited knowledge, I must conclude that the sketch is accurate.

Ted Phillips

Ted Phillips

[59] Berlitz, Charles, and William L Moore. 1997. *The Roswell Incident*. New York: Berkley Books.

Chapter 18: The Arroyo Photograph — A Close Examination

In recent years, a short video uploaded to the U.S. National Archives has circulated among Roswell researchers. Much of the footage is of limited relevance: it appears to show publishers or editors leafing through The Roswell Report: Fact vs. Fiction in the New Mexico Desert and briefly panning across shelves of books related to the Roswell incident. This portion of the video adds no new information to the case.

The significance of the upload lies elsewhere—specifically, in a single still image shown at the beginning of the video. At first glance, the photograph appears unremarkable: a barren arroyo landscape consistent with central New Mexico terrain. However, closer inspection reveals elements within the image that are difficult to immediately categorize.

Initial Observations

Two features in the photograph warrant closer attention.

On the left-hand side of the image is an object that appears metallic and reflective, contrasting with the surrounding terrain. When examined in higher resolution, its shape suggests a damaged structure rather than a natural rock formation. Some observers have noted features resembling a curved surface, a possible dome-like contour at one end, and what could be interpreted as a fin or raised structure at the opposite end. These impressions bear a superficial resemblance to craft descriptions given by certain

Roswell-era witnesses, including accounts attributed to individuals such as Dan Dwyer.

It is important to emphasize that such interpretations are not immediately obvious and are often dismissed at first as visual noise or coincidence. Indeed, several individuals consulted during this examination initially expressed skepticism, attributing the perceived structure to pareidolia—the human tendency to see meaningful patterns in ambiguous stimuli. Only after extended scrutiny did some acknowledge that the object's geometry appeared unusual, though still inconclusive.

A Second Anomalous Feature

To the right of the metallic-looking object, another form appears that some have interpreted as resembling a small body. When magnified, it presents a rough outline that could be read as a larger head relative to a smaller torso—features often cited in testimony describing alleged non-human entities. However, the distance, resolution, and lack of scale markers make such identification highly speculative.

At no point does the photograph provide sufficient clarity to confirm that this form represents a body, human or otherwise. Any such interpretation must therefore be treated as provisional and subjective.

Provenance and Context

The photograph itself is not new. It appears within The Roswell Report: Fact vs. Fiction in the New Mexico Desert, reproduced without detailed commentary or specific contextual explanation. The report does not identify the image as evidence of a craft, debris, or bodies, nor does it provide information regarding the date, photographer, or exact location beyond a general association with arroyo terrain. The absence of commentary is notable but not, by itself, evidence of concealment. Government reports frequently include illustrative images without exhaustive analysis, particularly when the images are not considered central to the report's conclusions. At the same time, the lack of clarification leaves the image open to interpretation and invites further scrutiny.

Limits of Interpretation

It would be inappropriate to describe this photograph as definitive proof of a crashed craft or recovered bodies. Single images—especially those lacking metadata, scale, and provenance—cannot bear that evidentiary weight. The image does not corroborate witness testimony on its own, nor does it independently verify claims made elsewhere in the Roswell record. What the photograph does represent is an unresolved visual anomaly. It contains elements that some observers find difficult to reconcile with purely natural explanations, while remaining insufficiently clear to support extraordinary conclusions. This places it in the same category as many Roswell-related materials: suggestive, incomplete, and resistant to final interpretation.

Reconstruction and Analytical Use

For the purposes of this study, enlarged and enhanced reconstructions have been produced based solely on what can reasonably be discerned from the image itself. These reconstructions are not presented as restorations of reality, but as analytical tools—intended to help readers understand why certain features have drawn attention and how those features compare with descriptions found in eyewitness testimony.

As with all visual reconstructions in this work, they are illustrative rather than evidentiary.

Conclusion

The Arroyo Photograph neither confirms nor disproves the Roswell body narrative. It does, however, illustrate a recurring problem in the case: the existence of materials that are officially acknowledged yet insufficiently explained. Whether the objects perceived in the image are debris, shadows, rocks, or something else entirely remains unresolved. What can be said with confidence is that the photograph does not close the Roswell case. Instead, like the Ramey Memo and various witness accounts, it occupies a liminal space—where certainty is absent, interpretation varies, and the historical record remains incomplete

The Arroyo photograph from the Roswell Report: Fact vs fiction in the New Mexico Desert (1994).

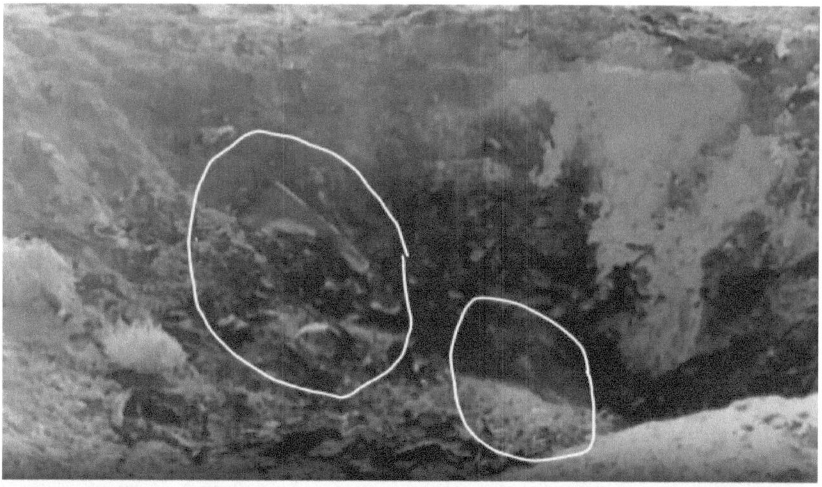

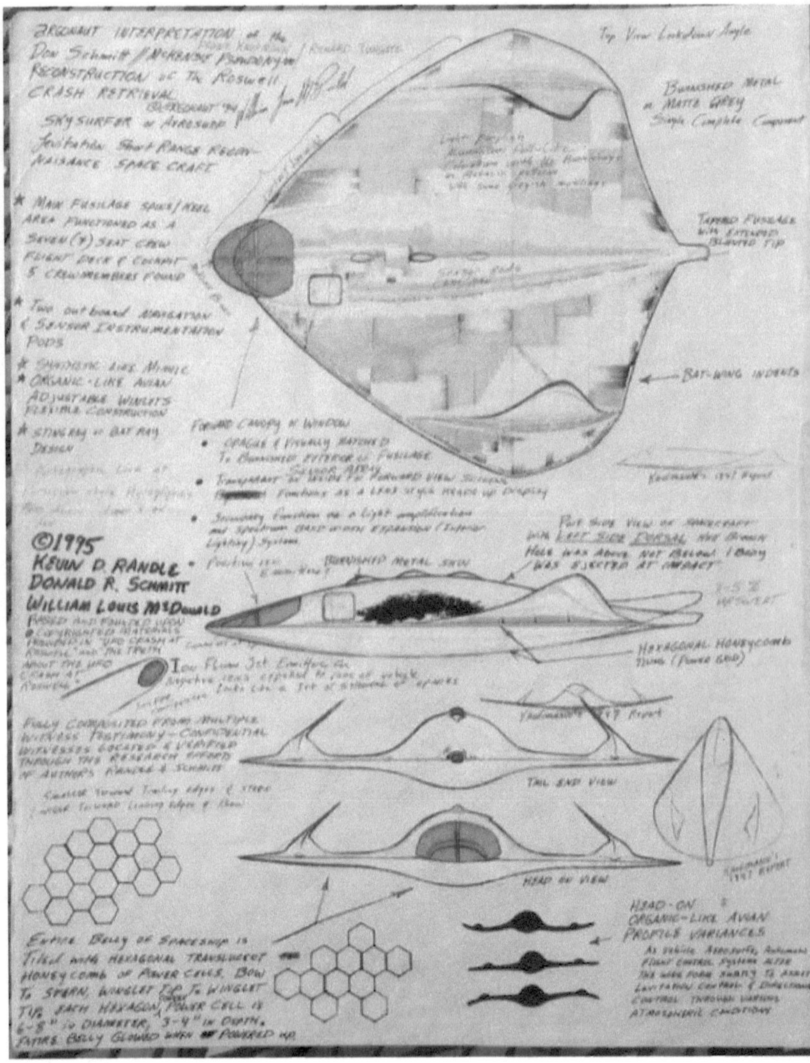

A sketch of the craft which bears striking resemblance to the photograph, reference to Kevin D Randle and Donald R Schmidt.

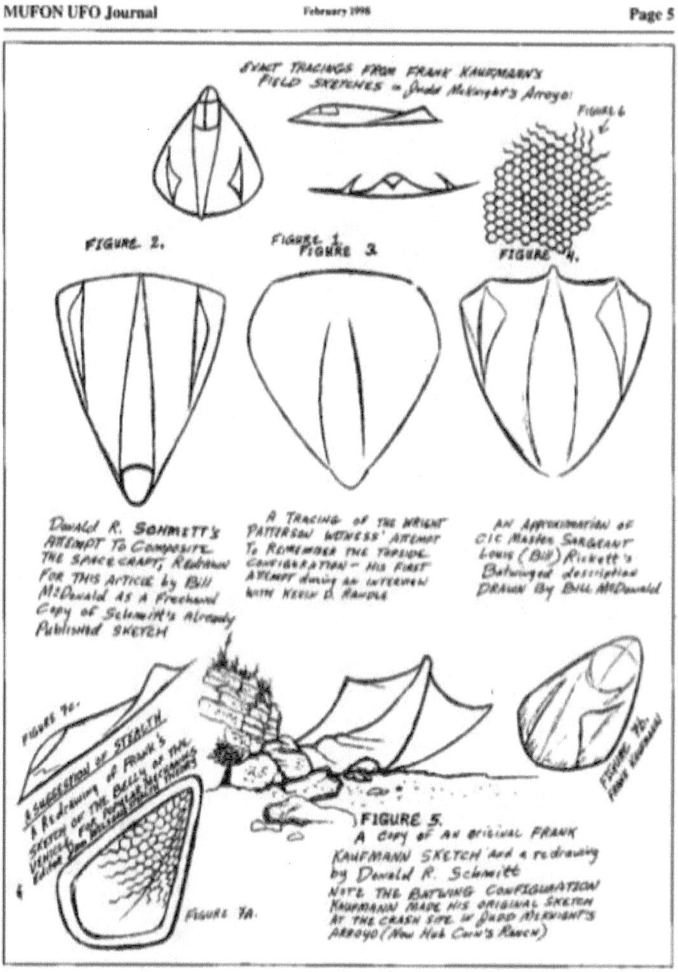

MUFON UFO Journal, sketch of the Roswell craft according to Eyewitness testimony.

The Arroyo photograph from the Roswell Report: Fact vs fiction in the New Mexico Desert with the craft and Alien body highlighted.

A close-up recreation of the craft in the arroyo.

A close-up recreation of the Alien body in the arroyo photograph based on eyewitness testimony.

The Arroyo Photograph: AI-Assisted Re-Examination and Author's Assessment

Following a renewed and enhanced digital rescan of the so-called "arroyo photograph," my assessment of this image has changed decisively. Earlier caution has given way not to speculation, but to sustained scrutiny. After repeated examination—both visually and through AI-assisted enhancement—I am now confident in stating that at least one object visible in the photograph does not belong naturally within the environment depicted. This conclusion was not reached quickly, nor casually. The image was examined for tonal continuity, spatial consistency, edge behaviour, and pattern repetition across multiple passes. What emerged was a set of uneven structural patterns that cannot be accounted for by erosion, shadowing, vegetation, or random geological debris. The spacing, angularity, and surface reflectivity of the object on the left side of the frame are anomalous when compared to the surrounding terrain.

In short: it does not behave like rock.

The object exhibits form characteristics inconsistent with the arroyo's natural morphology. Its outline shows abrupt transitions rather than gradual erosion. Its surface reflects light differently from adjacent material, suggesting a composition or finish distinct from the surrounding environment. These are not artefacts introduced by scanning or enhancement; they persist across contrast levels and processing methods. Equally important is the contextual absence surrounding the photograph. The image appears to function as b-roll—a general location shot included to provide environmental reference rather than to document a specific feature. This may explain why no descriptive caption or explanatory note accompanies it in official material. The photograph was treated as unimportant.

I believe that assumption was mistaken.

B-roll images are often included precisely because they are considered neutral and incidental. Yet it is sometimes within such material—unburdened by explanatory intent—that overlooked details persist. The absence of commentary does not indicate absence of significance; it indicates that significance was never evaluated.

To be clear: I am not arguing that this photograph, in isolation, constitutes definitive proof of an extra-terrestrial craft. That claim would exceed the evidentiary limits of a single image. What I am asserting—without reservation—is that the object identified through this rescan is not native to the environment shown and cannot be dismissed as a natural feature without explanation.

At this stage, skepticism requires more justification than confidence.

If the object is mundane, it must be accounted for: identified, contextualized, and explained within the physical realities of the location. To date, no such explanation has been offered. Until one is, the arroyo photograph remains not an illustration—but an unresolved data point.

And unresolved data points are where investigations begin, not end. Below you will find the re-enhanced AI scanned images.

AI enhanced picture with better of quality of the b-roll Arroyo photograph revealing the craft and body with more detail.

Closer up enhanced AI resolution of the b-roll Arroyo photograph close up.

Colour enhanced photograph using AI and colour imaging revealing the craft and body.

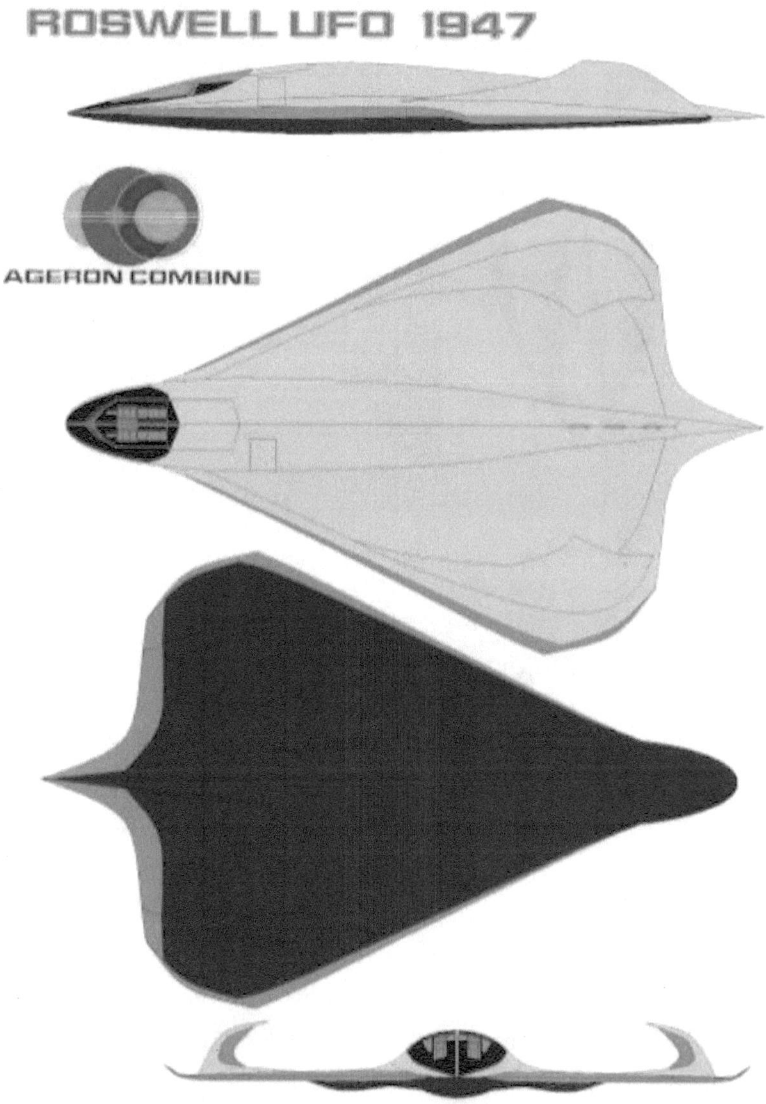
ROSWELL UFO 1947
AGERON COMBINE

[60]Modern illustrative reconstruction of a purported Roswell-type craft. This image is speculative and provided for comparative visualization only; it does not originate from U.S. military archives.

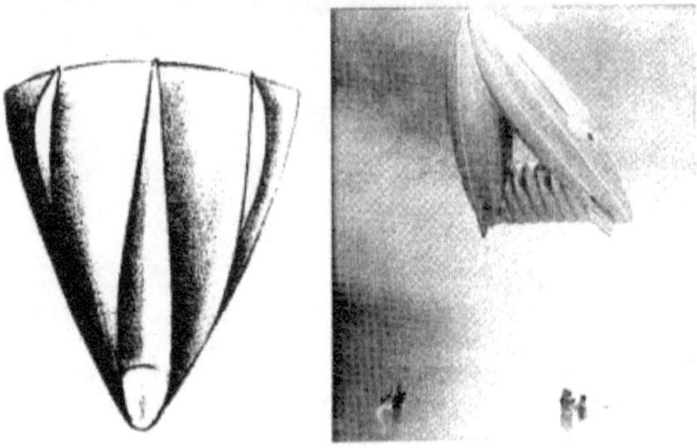

Fig. 56. *(Left)* A drawing from a popular UFO book, *The Truth About the UFO Crash at Roswell*, depicts an alien spacecraft allegedly drawn by an anonymous witness. *(The Truth About the UFO Crash at Roswell)*

Fig. 57. *(Right)* A tethered "Vee" balloon shown here at Holloman AFB, N.M. in March 1965. This experimental balloon, is strikingly similar to the "alien" craft. *(U.S. Air Force photo)*

[61]

The Air Force's own comparison images reveal an underlying methodological problem. Figure 56 reproduces a witness-derived drawing published decades after 1947, while Figure 57 depicts a tethered experimental balloon photographed at Holloman Air Force Base in 1965—eighteen years after the Roswell incident. The implication is not that the two objects are identical, but that later aerospace developments can be used retroactively to reinterpret earlier accounts. This approach does not resolve what witnesses described in 1947; it reframes those descriptions using technology that did not yet exist. As such, the comparison functions less as an explanation and more as a retrospective narrative control.

Chapter 19: The FBI Telexes: July–August 1947 and the Question of Official Concern

It is often asserted that the FBI played only a passive, clerical role in the flying disc wave of 1947—merely forwarding reports to the military and quickly disengaging. While it is true that the Bureau was not the lead investigative agency, declassified documents from July and August 1947 complicate this simplified narrative. They reveal an organization that was, at minimum, alert, internally communicative, and—at times—openly concerned about the security implications of the phenomenon.

The most frequently cited FBI document connected to Roswell is the 1950 memorandum authored by Guy Hottel and addressed to Director J. Edgar Hoover. That memo, discussed earlier in this book, is routinely dismissed as second-hand and uninvestigated. However, focusing solely on the Hottel memo obscures a broader documentary pattern: the FBI had already been receiving, indexing, and classifying flying disc reports as a *security matter* in the summer of 1947—weeks after the Roswell incident.

The August 11, 1947 FBI Teletype

One such document appears in *The Roswell Encyclopaedia* by Kevin D. Randle and is also reproduced in public archives via Project Gutenberg. Dated August 11, 1947, it is an FBI teletype marked **URGENT**, originating from the Portland Field Office and addressed directly to FBI Headquarters:

> *"FLYING DISCS. SECURITY MATTER..."*

The message concerns a former Navy pilot reporting sightings of a mysterious object while flying at 5,000 feet over Oregon. The language is notable not because of the sighting itself—many such reports existed—but because of the classification. Flying discs are explicitly labelled a *security matter*, not a curiosity, prank, or civilian misidentification. The teletype is recorded, indexed, and preserved.

This is inconsistent with later claims that the FBI regarded flying disc reports as trivial or unworthy of attention.

The July 8–9, 1947 Photograph Case

Another revealing document is an FBI memorandum dated August 8, 1947, referencing a photograph allegedly taken on **July 8, 1947**—the same day the Roswell Army Air Field issued its now-famous press release announcing the recovery of a "flying disc."

In this case, a Norfolk, Virginia schoolboy claimed to have photographed three fast-moving aerial objects, one large and two smaller, traveling at extreme altitude and speed. The FBI interviewed the witness, traced the photographic negative, documented its transfer to a commercial photo service, and noted profit-sharing arrangements. The report ends with a telling sentence:

> *"Inasmuch as the Army authorities... are cognizant of the above information, no further investigation will be conducted by this office."*

Here again, the FBI does not dismiss the report outright. Instead, it defers—explicitly—to military authorities, suggesting an established division of responsibility rather than disinterest.

The August 4, 1947 Hackensack Teletype

A similar pattern appears in an August 4, 1947 teletype from Newark regarding a sighting in Hackensack, New Jersey. Witnesses described an object "revolving slowly," moving rapidly, and explicitly stated it was neither a balloon nor a kite. The report was forwarded urgently, recorded, and indexed. Inquiry was ongoing.

Across these documents, a consistent behaviour emerges: the FBI logs sightings, evaluates credibility, forwards information through formal channels, and treats the subject as potentially relevant to national security.

The Schulgen–Hoover Memorandum: The Most Revealing Document

The most consequential document, however, is an internal FBI memorandum dated July 24, 1947—less than three weeks after the Roswell incident. It records a request from Brigadier General George F. Schulgen of Army Air Forces Intelligence for FBI cooperation on flying discs.

Director Hoover's handwritten response is unambiguous:

> *"I would do it but before agreeing to it we must insist upon full access to discs recovered. For instance in the La. case the Army grabbed it and would not let us have it for cursory examination."*

This sentence alone undermines the claim that no physical objects were ever recovered or that the FBI had no interest in examining them. Hoover's concern is not speculative sightings—it is *access to recovered material.* General Schulgen subsequently assures the FBI that all recovered discs will be made available for examination and that scientific findings will be shared.

Equally revealing is Schulgen's acknowledgment that the matter should be followed through to determine whether the discs were real and to establish their origin—paired with his suspicion that some reports might involve subversive activity or mass hysteria. This reflects a dual concern: physical reality on one hand, psychological or political manipulation on the other.

Conclusion: Interest Without Answers

The FBI's role in 1947 was limited—but it was not indifferent. The Bureau's documents reveal an institution navigating unfamiliar terrain, deferring authority while quietly insisting on access, classification, and control. If nothing unusual had occurred—if there were no recoveries, no materials, no reason for concern—then Hoover's insistence on examining recovered discs makes little sense.

What remains unresolved is not whether the FBI knew the answers, but whether it was ever allowed to see them.

Chapter 20: Marcel's Diary and the Question of Psychological Impact

If, as some accounts suggest, Jesse A. Marcel was directly involved in the recovery of an anomalous craft in July 1947, it would be reasonable to expect some trace of psychological disturbance, fixation, or emotional strain to appear in his private writings shortly afterward. Traumatic or extraordinary experiences often leave marks—not necessarily in explicit confession, but in tone, preoccupation, or disorganization. For this reason, the August 1947 diary page attributed to Marcel warrants careful consideration. What is striking about the page is not what it contains, but what it lacks. The writing consists of detached aphorisms—philosophical observations on time, age, and human relationships—expressed in a calm, almost literary manner. There is no indication of agitation, obsession, fear, or rupture from ordinary thought. If the page is authentic and contemporaneous, it suggests a man who remained composed, reflective, and psychologically intact in the weeks following the Roswell incident.

This does not disprove Marcel's later claims, nor does it confirm them. Instead, it challenges a common assumption: that proximity to an extraordinary event must necessarily produce visible psychological disturbance. The diary implies the opposite—that if Marcel was exposed to something profoundly unusual, he absorbed it without collapse, hysteria, or apparent fixation. In that sense, the document may be read not as evidence against his involvement, but as a testament to his temperament. Marcel was a career intelligence officer, trained to compartmentalize information and operate under pressure. His service record suggests familiarity with secrecy, ambiguity, and high-stakes responsibility. The absence of emotional leakage in his private writing may therefore reflect professional discipline rather than ignorance. If so, the diary page reinforces a portrait of Marcel as a man capable of witnessing extraordinary events without allowing them to dominate his inner life.

Ultimately, the diary does not resolve the Roswell question. But it does narrow the range of plausible interpretations. If Marcel was involved, he did not unravel. If he knew something, it did not unseat him. And if nothing unusual occurred, the diary is exactly what one would expect from a reflective, orderly mind. Either way, the page argues against caricature—against the image of Marcel as either a fantasist or a broken man—and supports a more nuanced understanding of his character. Below I have included the diary page of Jesse A Marcel from August of 1947, which would have been a month after the Roswell crash. We know Marcel would have still been working at the RAAF, Roswell Army Airfield nothing in his life had changed dramatically after the event. This is the diary page that is currently in dispute by historians and scholars.

"A beautiful woman
is the paradise of the eyes,
the hell of the soul, and the
purgatory of the purse."
Time and tide wait
for no man — but time
always stands still for
a woman of thirty."

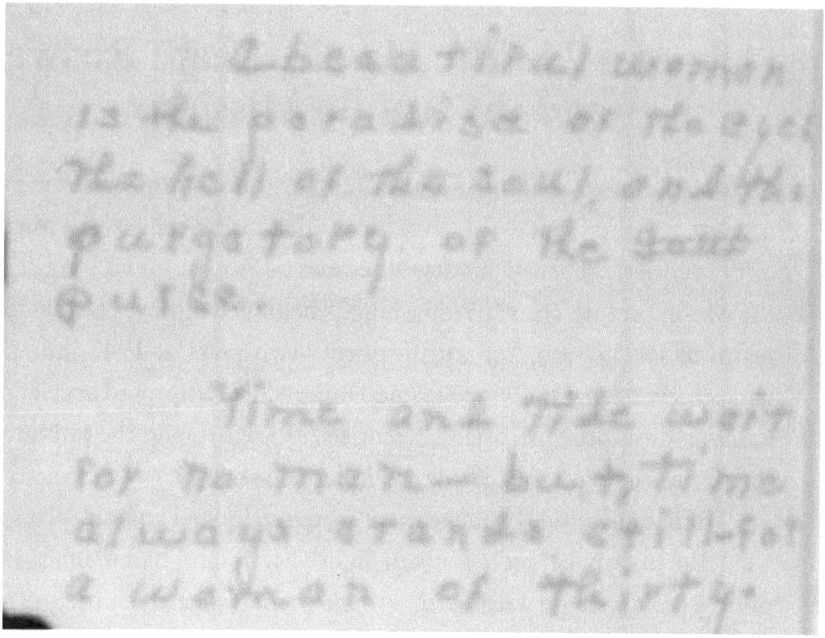

A beautiful woman is the paradise of the eyes, the hell of the soul, and the purgatory of the purse.

Time and tide wait for no man—but time always stands still for a woman of thirty.

Chapter 21: The Massive Leap: Roswell and the Apollo Missions

One of the most striking—and least questioned—aspects of modern history is the speed of America's technological ascent in the mid-twentieth century. Just two decades before Apollo 11 landed on the Moon on July 16, 1969, spaceflight belonged almost entirely to the realm of speculation. Yet within twenty-two years of 1947, humanity had achieved what had once seemed impossible. This rapid acceleration has long prompted historians, scientists, ufologists, and the public alike to ask the same question: how did it happen so fast?

To that question, I believe there is an overlooked answer—Roswell.

I do not make such a claim lightly. It is not based on a single photograph, a single witness, or a single theory. Rather, it emerges from a convergence of testimony, technological outcomes, and historical context that, taken together, suggest an anomalous catalyst at work beneath the official narrative of steady progress. At first glance, the Apollo 11 Command Module and the alleged Roswell craft appear to share little in common. But closer examination reveals an unexpected parallel. When one studies the underside of the Apollo Command Module, the heat shield is revealed as a precisely engineered honeycomb structure—hexagonal cells forming a resilient, ablative barrier designed to endure the most extreme thermal stresses imaginable.

Convergence or Continuity? Roswell and the Architecture of Apollo

By the time Apollo spacecraft began returning safely through Earth's atmosphere, their heat shields represented the pinnacle of human engineering. The ablative underbelly—constructed around a honeycomb geometry—was capable of withstanding temperatures that would reduce conventional metals to vapour. Officially, this achievement is explained as the result of incremental research, rooted in pre-war materials science and refined through years of testing. That explanation is tidy. It is also incomplete. The issue is not whether ablative heat-shield concepts existed before 1947—they did. The deeper question is whether the speed, confidence, and decisiveness with which such technology matured in the decades following Roswell can be fully explained without acknowledging an anomalous influence.

Witnesses to the Roswell debris consistently described materials that defied their expectations: lightweight yet structurally resilient; foil-like yet unburnable; flexible yet capable of returning to their original shape. Some accounts—often dismissed as exaggeration—refer to layered or structured materials that did not behave like conventional metals known at the time. When later composite drawings and reconstructions of the alleged craft are examined, a recurring feature appears: a segmented or patterned underside, suggestive not of smooth aerodynamic skin, but of engineered thermal or structural complexity.

It is here that the comparison to Apollo becomes unavoidable.

The honeycomb architecture of the Apollo heat shield was not merely an aesthetic or conventional design choice. It was a solution to extreme constraints: heat, weight, stress, and survivability. It is precisely the type of solution one would expect if engineers were attempting to replicate, adapt, or surpass material behaviours that exceeded known limits. The resemblance between the Apollo underbelly and Roswell-era reconstructions may be dismissed as coincidence, but coincidence becomes less persuasive when it repeats across eyewitness testimony, imagery, and technological outcome. Critics correctly note that honeycomb structures are common in engineering. That is true. What is less often addressed is why such structures became central, rapidly, and decisively in post-war aerospace development—and why materials described in 1947 appeared, to contemporary observers, so far ahead of their time.

The possibility that Roswell introduced U.S. military research divisions to unfamiliar material behaviours is not fringe speculation. It is consistent with how classified research environments operate. Colonel Philip J. Corso—whose claims remain controversial but influential—asserted that recovered materials informed advances in fiber optics, integrated circuits, and Kevlar. Whether one accepts Corso's account in full or in part, the broader principle remains sound: reverse engineering does not require complete understanding. It requires fragments, performance benchmarks, and observed behaviours. NASA, formally established in 1958, inherited personnel, facilities, and classified research streams from military and intelligence communities already grappling with unconventional aeronautical challenges. It is implausible to assume that if anomalous materials had been studied under military auspices, their influence would remain permanently isolated—especially as spaceflight became the new strategic frontier. NASA's silence regarding any Roswell-related influence does not resolve the issue; it reflects an institutional continuity shaped by Cold War secrecy and compartmentalization.

This is not an argument that Apollo spacecraft were alien machines. They were unquestionably human-built. It is, however, an argument that the trajectory of their development may have been subtly altered—nudged forward—by exposure to materials or concepts that did not originate within conventional human design pathways. In that light, the similarities between Roswell craft reconstructions and the Apollo heat shield are not proof, but neither are they trivial. They suggest a continuity of problem-solving that may have begun earlier than officially acknowledged. When eyewitness testimony, composite imagery, and technological outcomes align too closely, coincidence becomes a weaker explanation than convergence guided by discovery. NASA may never speak openly about such influence. If Roswell involved something genuinely unknown, silence would be the most predictable response—not admission. History shows that revolutionary technologies are rarely unveiled with full disclosure of their origins. They arrive sanitized, domesticated, and attributed to steady progress.

Perhaps Apollo was exactly that: steady progress. Or perhaps it was steady progress quietly informed by a moment in the New Mexico desert, when engineers first encountered materials that behaved as though they belonged to a different technological order. The historical record does not allow certainty. But it does allow suspicion—and in historical inquiry, sustained suspicion is often the first indication that a story has not yet been fully told.

[62]Apollo Command Module (Museum Display)

Apollo Command Module displayed with Earth backdrop (Smithsonian/NASM-style exhibit photograph).

This image represents the Apollo Command Module as preserved and displayed by NASA and the Smithsonian, emphasizing the spacecraft's external structure and heat-shielded lower section following atmospheric re-entry.

[63]

Apollo Command Module Heat Shield – Close-Up (Ablative Pattern)

Close-up photograph of the Apollo Command Module's ablative heat shield showing circular and hexagonal ablation patterns after re-entry.

The honeycomb-like and cellular ablation pattern is the result of the heat shield's phenolic epoxy resin system, designed to char and erode in a controlled manner during re-entry.

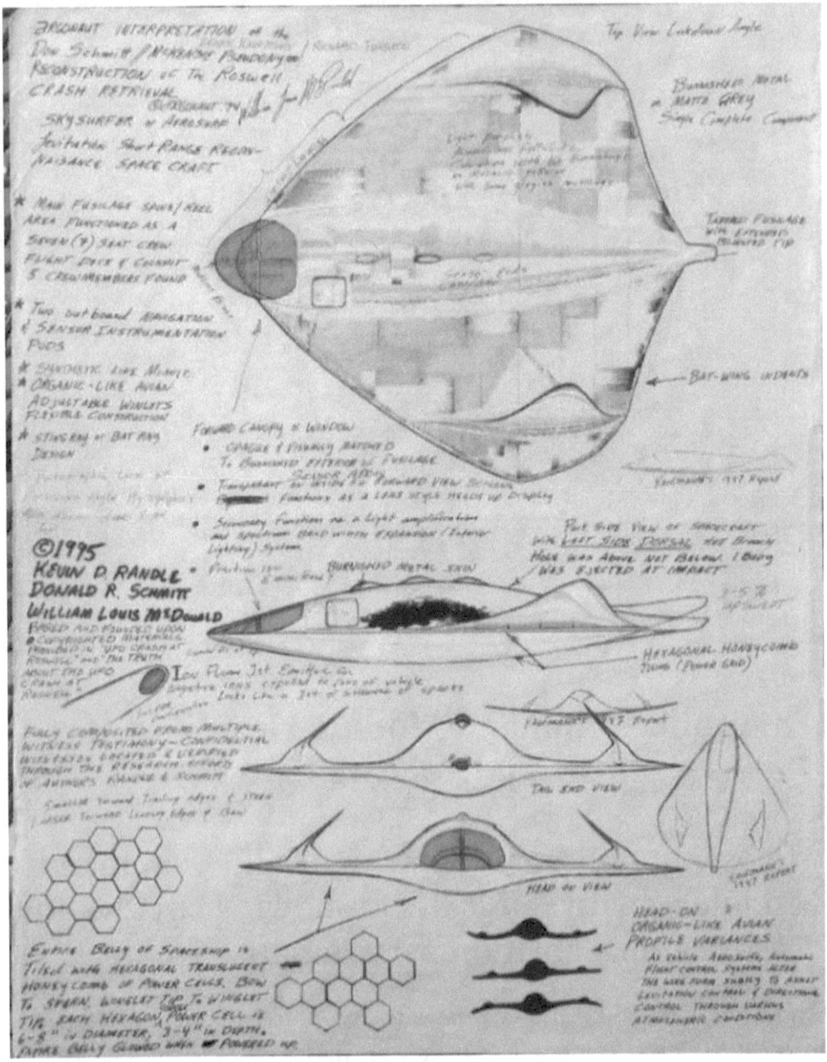

[64] Roswell Craft Reconstruction Drawing (Kevin D. Randle et al.)

Annotated composite reconstruction drawing of a Roswell craft based on multiple eyewitness testimonies.

Composite Roswell Craft Reconstruction Drawing, c. 1990s.

Reproduced in:

Kevin D. Randle, The Roswell Encyclopaedia (1999)

Kevin D. Randle & Donald R. Schmitt, The Truth About the UFO Crash at Roswell (1991)

This reconstruction reflects synthesized eyewitness descriptions rather than physical artefacts. Notably, it depicts a segmented or hexagonal underside, a feature repeatedly mentioned in testimony and later compared by researchers to advanced aerospace structural designs

Chapter 22: The UFO Crash at Aztec, New Mexico

An incident that has long existed at the margins of UFO history—but has only recently begun to receive renewed attention from serious investigators—is the reported UFO crash near Aztec, New Mexico. For decades, the Aztec case was largely dismissed, not necessarily because it lacked witnesses or internal consistency, but because it became entangled early on with discredited sources. That stigma, once attached, proved difficult to remove. The Aztec story entered public consciousness in 1950 with the publication of Behind the Flying Saucers by journalist Frank Scully. In his book, Scully claimed he had received information from a scientist he referred to as "Dr. Gee," described as a German-trained researcher with an impressive academic background. This individual was later identified as Leo Gebauer, who, along with Silas Newton, provided Scully with the account that would define the Aztec incident.

According to Scully's sources, the crash occurred in March 1948, near Aztec, New Mexico, close to the Colorado border. The object was described as a large disc-shaped craft, significantly larger than the one associated with Roswell the year before. Inside, they claimed, were sixteen humanoid occupants, all reportedly deceased. The recovery operation was said to involve U.S. military personnel, as well as scientific teams, and the craft itself was allegedly transported to Wright Field (later Wright-Patterson Air Force Base)—the same facility frequently linked to the Roswell debris. The credibility of Scully's account suffered a major blow when Newton and Gebauer were later exposed as fraudsters in an unrelated oil-swindle scheme. Although their convictions had nothing to do with UFOs, the damage was done. The Aztec story became synonymous with deception, and for decades it was treated as a cautionary tale—used to discredit not only the incident itself, but UFO research more broadly.

Yet the dismissal of the Aztec case may have been premature.

In 2015, nuclear physicist and veteran UFO researcher Stanton T. Friedman, alongside Scott Ramsey, revisited the incident in their book The Aztec UFO Incident: The Case, Evidence, and Elaborate Cover-Up of One of the Most Perplexing Crashes in History. Their work did not rely on Newton or Gebauer as primary witnesses. Instead, it focused on independent testimony, local accounts, and documentary traces that had been largely ignored or overshadowed by the earlier scandal. Ramsey and Friedman documented accounts from a local rancher and oil field workers who claimed to have arrived at a remote worksite only to discover a crashed saucer already present. According to these witnesses, the craft was partially intact. Inside, they observed two humanoid figures slumped over a control panel, their bodies badly charred. The windows of the craft were described as having properties similar to photochromic glass, darkening or tinting in response to sunlight—an unfamiliar technology to observers in the late 1940s.

Witnesses also recalled the presence of a Bell H-1 military helicopter circling the area—an aircraft that, at the time, was itself cutting-edge technology. Several noted that this was the first time they had ever seen such a helicopter, adding to their sense that the situation was extraordinary and tightly controlled. What makes the Aztec case compelling is not that it mirrors Roswell in every detail—it does not—but that it appears to share structural similarities: rapid military response, removal of material, restricted access, and the quiet disappearance of physical evidence. Unlike Roswell, however, Aztec was burdened early by compromised messengers. When those messengers fell, the message was discarded with them. This chapter does not argue that every claim made by Frank Scully was accurate, nor does it attempt to rehabilitate unreliable sources. Instead, it asks a narrower and more important question: was a genuine incident obscured by the involvement of the wrong intermediaries? If so, Aztec may represent not a fabricated legend, but a second, poorly handled chapter in the early history of unidentified aerial phenomena. The Aztec crash, long dismissed, may yet warrant a place alongside Roswell—not as proof, but as a reminder that truth and falsehood often travel together, and that history sometimes discards the former because of the latter.

Alleged Military Containment and Cover-Up at Aztec

In *The Aztec UFO Incident*, Chapter 4 focuses on what Friedman describes as a deliberate and more sophisticated military response to the Aztec crash—one that appears to reflect lessons learned from Roswell just three years earlier.

According to Friedman, the U.S. Army's Fifth Army was allegedly deployed to the Aztec site shortly after the crash was discovered. Estimates cited in the book suggest approximately 200 military personnel were involved in the recovery, security, and site remediation phases. Unlike Roswell, where information briefly escaped through a public press release, the Aztec response was reportedly sealed from the outset.

Securing and Restructuring the Crash Site

Friedman describes accounts in which the crash area was physically altered after recovery operations. Heavy equipment was allegedly used to:

Remove all visible debris

Regrade or reshape the terrain

Eliminate scorch marks or impact indicators

Restore vegetation or natural ground contours

The stated purpose, according to witnesses cited in the book, was to ensure that any later visitor would find no evidence that an object had ever crashed there.

This represents a key distinction from Roswell:
Roswell was cleaned up quickly, but imperfectly.
Aztec, Friedman argues, was erased.

Controlled Access and Silence

The book further claims that:

1. Civilians were removed from the area
2. Workers and witnesses were warned not to speak
3. Military personnel involved were compartmentalized
4. No local press statements were issued

Friedman explicitly contrasts this with the Roswell incident, where the now-famous "flying disc" press release demonstrated a failure of message control. At Aztec, he argues, no such failure occurred.

Learning from Roswell

A central thesis of Chapter 4 is that Aztec reflects institutional learning.

Friedman suggests that by 1950:
The military understood the public impact of Roswell
Information leaks were seen as strategic failures
UFO recovery protocols had become more refined
Aztec, in this framework, was handled as a black operation, not a local anomaly.

> Roswell showed what *not* to do.
> Aztec showed what happens when secrecy is enforced from the start.

Why Aztec Was Discredited

Chapter 4 also addresses why the Aztec case was later dismissed so aggressively. Friedman argues that:

1. The criminal convictions of Silas Newton and Leo Gebauer provided a convenient way to invalidate the entire case
2. Discrediting the messengers allowed the event itself to be erased from serious consideration
3. The stigma attached to Aztec discouraged researchers for decades

Friedman's position is not that every claim made by Scully or Newton was true, but that the presence of unreliable intermediaries does not automatically negate the possibility of a real event—especially when later witness testimony and patterns of military behaviour align with known secrecy practices.

Dr. "G," Pseudonyms, and the Architecture of Discredit

An additional layer of complexity in the Aztec case lies in the identity of the figure Frank Scully referred to as Dr. "G." Subsequent research has suggested that "Dr. Gebauer" may not have referred to a single individual at all, but rather functioned as a pseudonym or composite identifier for multiple scientists involved in classified research. Some researchers have proposed that as many as nine individuals may have been represented under this single designation, a practice not unheard of in highly compartmentalized programs where deniability was essential. Within Behind the Flying Saucers, Scully recounts descriptions attributed to Dr. G that are undeniably strange—yet technologically intriguing. Among these is mention of a small, box-like device, roughly the size of a pack of king-sized cigarettes, which allegedly emitted a pulsed frequency at regular intervals. Accounts vary on the exact timing—whether every fifteen minutes per hour or fifteen seconds per hour—but the emphasis is not on precision so much as intent: the suggestion of a compact, autonomous signalling or stabilizing technology.

Perhaps more compelling is Scully's account of the photographic limitations imposed at the Aztec recovery. He wrote that images of the craft were captured using a specialized film that would self-degrade within approximately two hours of light exposure, and that even when chemically treated, the developed images would survive only briefly before deteriorating. Whether literal or metaphorical, this description speaks directly to a recurring problem in UFO history: the systematic absence of durable physical documentation, even when witnesses insist such documentation once existed. It is important to be clear: accepting that Scully relayed these claims does not require accepting them as factual. What matters is why stories like this repeatedly emerge, and why they are so consistently rendered unverifiable. Frank Scully himself was not a fringe figure. He was a respected journalist, a man of considerable education, and someone with access to scientific and political circles. The collapse of his Aztec narrative did not occur because the claims were disproven through direct examination of the evidence, but because the sources who provided him the information were later exposed as fraudsters in an unrelated oil-swindle case. Once that happened, the entire account became radioactive—not merely questionable, but untouchable.

And this is where the pattern becomes impossible to ignore.

If Scully's account had been accurate, it would have been nearly impossible to prove. The alleged physical evidence was ephemeral by design. The witnesses were second-hand. The primary informants were later discredited. The institutions implicated—the military, intelligence services, and scientific contractors—were precisely those with both the means and precedent to reshape narratives, expose damaging personal histories, and redirect public attention. This does not mean that Silas Newton and Leo Gebauer were telling the truth. But it does raise a more unsettling question: what if they were telling a version of it? And if so, how easy would it have been to ensure that no version of their story could ever be taken seriously again? This is not unique to Aztec. It is the same structural problem that appears at Roswell and echoes through nearly every major UFO incident that followed. Physical evidence is absent or compromised. Witness credibility is undermined years later. Official explanations arrive long after the fact, confident not because they resolve every inconsistency, but because alternative explanations have already been neutralized.

There is never a smoking gun—only a trail of testimonies, recantations, reputational collapses, and explanations that close cases without fully answering them. That recurring absence may itself be the point. Whether Aztec represents a genuine crash, an elaborate fabrication, or a hybrid of truth and misdirection, it fits seamlessly into a broader historical pattern: one in which extraordinary claims are not disproven so much as rendered impossible to prove. And in that sense, the Aztec case does not stand apart from Roswell—it reinforces it

The Discrediting of Aztec: Motive and Method

One of the most revealing—and often overlooked—aspects of the Aztec UFO incident is not the testimony itself, but the manner in which it was dismantled. Central to this effort was journalist J. P. Cahn, whose work did less to disprove the alleged crash than to discredit the individuals associated with its disclosure—most notably Frank Scully. Cahn's campaign focused almost exclusively on exposing the past business dealings of Silas Newton and Leo Gebauer, portraying them as con men and, by extension, rendering the Aztec story fraudulent by association. Yet this approach raises an important question: was the incident disproven, or merely buried beneath the reputational ruin of its sources?

The oil industry of the late 1940s and early 1950s was rife with speculative ventures, exaggerated claims, and legal gray areas. Practices that later resulted in fraud convictions were not uncommon at the time, and in some cases were selectively prosecuted. While this does not absolve Newton or Gebauer of wrongdoing, it complicates the narrative that their legal troubles alone invalidate every claim attributed to them. More striking still is the fate of Frank Scully himself. A respected journalist with no history of fraud or deception, Scully became collateral damage in the effort to dismantle the Aztec story. Rather than addressing the substance of the claims he reported, critics focused on undermining his credibility through association. Once Scully was discredited, the incident could be dismissed without further examination.

This pattern—discredit the witnesses, attack the messenger, and avoid addressing the event itself—has repeated throughout UFO history. It does not prove that the Aztec crash occurred, but it does suggest that the mechanisms used to suppress the story deserve scrutiny in their own right. When a narrative is destroyed through character assassination rather than evidentiary rebuttal, historians are right to ask whether the full truth has ever been allowed to surface.

The Farmington Armada (March 1950): Why It Matters More Than It Is Treated

On March 17, 1950, the Farmington Daily Times ran a front-page headline that stunned both the town and later UFO researchers:

"Huge Saucer Armada Jolts Farmington"

Unlike many UFO reports of the era, this was not a single witness sighting, nor a fleeting observation. It was a mass sighting, reported across the town over the course of the day, involving hundreds of witnesses from different backgrounds—housewives, businessmen, students, and local officials. Two years after the alleged Aztec crash, the nearby town of Farmington, New Mexico became the site of one of the most significant mass UFO sightings in American history. On March 17, 1950, hundreds of witnesses reported observing numerous disc-shaped objects flying in coordinated formations over the town in broad daylight. The event, documented by the Farmington Daily Times, involved manoeuvring behaviour—sudden turns, speed changes, and formation flying—that bore no resemblance to known aircraft of the period. Despite its scale, the incident was quickly relegated to the margins of UFO history. Its proximity to Aztec, however, invites closer examination. While no direct causal link can be established, the Farmington sighting reinforces the conclusion that northern New Mexico was a sustained focal point of anomalous aerial activity in the late 1940s and early 1950s, rather than the location of a single isolated event.

Among the most credible witnesses to the Farmington sightings was Clayton J. "Buddy" Body, a former U.S. Air Force serviceman. Body testified that the objects appeared to be flying at an altitude of approximately 15,000 feet and estimated their number at nearly 500. He described the craft as moving in coordinated formations, exhibiting controlled manoeuvrability inconsistent with known aircraft or atmospheric phenomena. Body's military background is significant, as it suggests familiarity with aircraft behaviour, altitude estimation, and flight dynamics. His testimony reinforces the conclusion that the Farmington incident was not a case of isolated misidentification, but a large-scale, sustained aerial event observed by experienced and untrained witnesses alike.

WWII Tail Gunner Testimony – Movement Beyond Aircraft Capability

Also among the witnesses to the Farmington sightings was a World War II veteran who had served as a tail gunner, a role that demanded constant situational awareness, aircraft recognition, and the ability to judge speed, distance, and relative motion under combat conditions. His testimony is especially valuable because it was grounded not in speculation, but in professional experience.

The witness stated that the movement of the objects immediately ruled out conventional aircraft. He described their motion as unlike anything he had encountered during the war—neither fighter planes, bombers, nor experimental aircraft. In particular, he emphasized their ability to change direction smoothly and rapidly without banking, losing altitude, or exhibiting the inertia expected of known flying machines.

When military authorities later suggested that the objects might have been "cotton wool" or airborne debris, the witness rejected the explanation outright. He stated plainly that such material would drift irregularly, lack cohesion, and could not manoeuvre in coordinated patterns at altitude. As someone trained to distinguish between aircraft, flak, cloud formations, and battlefield debris, he maintained that he knew exactly what he had observed.

His conclusion was simple and firm: whatever the objects were, they did not behave like aircraft—nor like anything naturally carried by the wind.

FARMINGTON DAILY TIMES

HUGE 'SAUCER' ARMADA JOLTS FARMINGTON

Jury Gets Bribe Case; Heard Denies Charges

Veracity of Negro Attacked

Crafts Seen By Hudreds

Blackout Game Preceded Slaying Of Pretty Co-ed

Newsman Writes Letter To Flying Saucer Pilot

Drafting Reply To Commie Charge

Entire Crew Rescued From Exploded Ship

Scores Report Seeing Saucers' Flight in Formation Over New Mexico

FARMINGTON, N.M., March 17 (U.P.)—A mass flight of flying saucers over this Northwestern New Mexico oil town of 5000 population was reported today by 50 persons, including businessmen and private pilots.

Detailed accounts of the flight varied, but the witnesses estimated the number of the objects, which appeared in groups commencing at 10:30 a.m., and lasting for one hour, from one to "hundreds."

The advertising manager of the Farmington Daily Times, Clayton Boddy, said he was with a group of five businessmen who saw a mass flight of the saucers. Boddy said they were all disk-shaped and fluttered through the air without descending to earth.

Among the saucers spotted by scores of persons on Farmington's streets was one low-flying, red-hued, saucer-shaped object which streaked across the sky.

Orville Ricketts, editor of the Times, said a cross check of witnesses revealed that the object cleared the horizon in its tremendous speed in about three seconds.

Ricketts said he did not see the objects, but that Boddy said that in the mass flight of the saucers he and five businessmen with whom he was standing believed they saw more than 100 of the strange objects.

The town's business was brought practically to a standstill while persons scanned the skies.

All of the objects were of a silvery color except for the one red "saucer," and all except the latter appeared to be at an extremely high altitude, Ricketts said.

Ricketts said that among the witnesses were airplane pilots.

Los Angeles Times 3/18/50

1. Farmington Daily Times — "Huge 'Saucer' Armada Jolts Farmington"

Primary Reference

Farmington Daily Times,

"Huge 'Saucer' Armada Jolts Farmington",

Farmington, New Mexico, March 17, 1950, front page.

Context for Your Book

This is the original local newspaper report

Based on dozens to hundreds of eyewitnesses

Reports objects flying in formation, some manoeuvring at low altitude

Describes public disruption and widespread observation across the town

Archival Availability

Farmington Daily Times archives

Microfilm collections (Library of Congress)

Reproduced in multiple UFO research works, including:

Kevin D. Randle

Stanton T. Friedman

NICAP historical archives

2. Los Angeles Times — "Scores Report Seeing Saucers' Flight in Formation Over New Mexico"

Primary Reference

Los Angeles Times,

"Scores Report Seeing Saucers' Flight in Formation Over New Mexico",

Los Angeles, California, March 18, 1950, p. A3.

(United Press wire service report)

Context for Your Book

Confirms the Farmington event as a national news story, not a local anomaly

Attributes testimony to:

Clayton J. "Buddy" Body (advertising manager, Farmington Daily Times)

Businessmen, pilots, and civilians

Reports estimates ranging from dozens to hundreds of objects

Notes extreme altitude and speed inconsistent with aircraft of the era.

Supplementary Notes: Military Involvement and Technological Context

One of the most compelling aspects of the Aztec, New Mexico case is the reported involvement of the U.S. Army's Fifth Army, alongside the alleged use of Bell helicopters during recovery operations. This detail is particularly noteworthy given that the Bell H-13 helicopter—often referenced in Aztec accounts—had only recently been patented and introduced in the late 1940s. If Bell helicopters were indeed deployed to the Aztec site as early as March 1948, this would place advanced rotary-wing aircraft in operational use at a remarkably early stage. Confirmation of such deployment would significantly strengthen the credibility of the Aztec account and warrants further archival investigation.

Equally important is the broader technological environment of the late 1940s and early 1950s in New Mexico. During this period, the region served as a testing ground for a wide range of experimental and classified aircraft and weapons systems, many of which were unknown to the civilian population. These included V-2 rockets, captured German ME-262 jet technology, and early experimental aircraft programs conducted in near-total secrecy. As a result, local witnesses were already accustomed to seeing unconventional and often startling aerospace technology. That context is crucial: it suggests that reports of a disc-shaped craft were not the product of naïveté or technological unfamiliarity. On the contrary, observers in the region were arguably better positioned than most civilians to distinguish between experimental military hardware and something truly anomalous. This context also raises a critical point: if experienced observers—accustomed to cutting-edge and bizarre aircraft—still identified the Aztec object as fundamentally different, then their testimony carries added weight.

Finally, it is worth noting that the U.S. military's later interest in disc-shaped aircraft is well documented. Declassified records confirm that by 1952, the Avro Canada VZ-9AV Avrocar project was receiving U.S. funding. Although officially framed as a vertical take-off experimental craft, its saucer-like design invites deeper scrutiny. The Canadian government ultimately withdrew from the project due to escalating costs, but the United States continued to pursue it for a time. This naturally raises an unresolved question: what earlier design concepts or recovered data influenced the Avrocar's form? That question will be explored in greater detail in Chapter 26. Taken together, these factors—early helicopter deployment, large-scale military involvement, a region saturated with classified testing, and later disc-shaped aerospace projects—form a coherent pattern. Whether examined individually or collectively, they suggest that the Aztec incident cannot be dismissed as a simple fabrication. Instead, it occupies a critical and still-unresolved position within the broader history of post-war military secrecy and unidentified aerial phenomena.

Composite Illustrations and Map Associated with the Aztec UFO Incident

Description & Context

This composite image brings together several elements associated with the **Aztec, New Mexico UFO crash narrative**, including:

Hand-drawn sketches of alleged occupants and interior layouts of a disc-shaped craft

A schematic representation of a circular vehicle

A directional map providing instructions to the purported Aztec crash site

Such illustrations have appeared in multiple works discussing the Aztec case, particularly those that rely on **second-hand witness testimony**, recollections from oil-field workers, and later reconstructions by researchers. The drawings are not presented as contemporaneous technical schematics, but as **retrospective visualizations** intended to synthesize multiple verbal accounts into a coherent image.

The inclusion of a detailed map is especially notable. It reflects the efforts of later investigators—most prominently **Stanton T. Friedman and Scott Ramsey**—to treat the Aztec case as a recoverable historical event rather than mere folklore. The map underscores that the story was tied to a **specific, inspectable geographic location**, rather than an abstract or shifting narrative.

Within this book, the image is used to illustrate how the Aztec incident developed a **parallel evidentiary structure** to Roswell: witness drawings, interior descriptions, humanoid figures, and claims of military recovery—followed by systematic discrediting of sources rather than definitive disproof of the event itself.

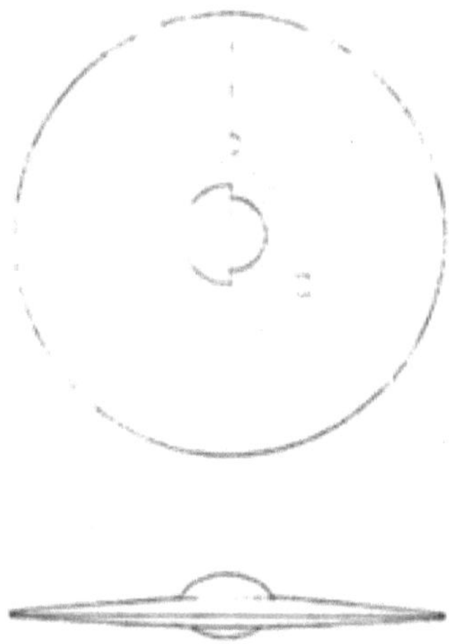

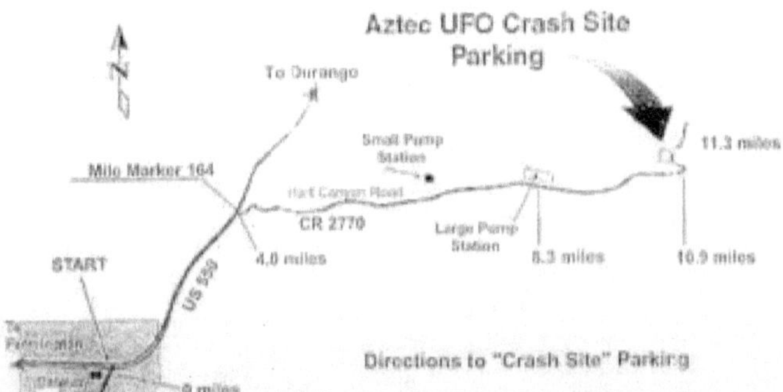

Directions to "Crash Site" Parking

[65]**Friedman, Stanton T., and Scott Ramsey. *The Aztec UFO Incident*. New Page Books, 2013.**

Scully, Frank. *Behind the Flying Saucers*. Henry Holt & Company, 1950

.Ramsey, Scott. Field investigation materials and site documentation cited in *The Aztec UFO Incident*.

Chapter 23: The inside of the craft a thorough examination

Philip J. Corso — The Day After Roswell (1997)

Corso did not claim first-hand entry into the craft, but stated that he was briefed on materials, reports, and artefacts derived from Roswell while working at the Pentagon (Foreign Technology Desk).

Interior Layout (as described)

Extremely compact interior

No standing room for adult humans

Proportions suggested occupants approx. 3–4 feet tall

Smooth, seamless surfaces

No visible rivets, bolts, or fasteners

Walls appeared grown or moulded, not assembled

Curved geometry

No sharp corners

Interior surfaces flowed continuously into one another

Controls & Instrumentation

No recognizable switches, dials, or gauges

Control interface described as integrated into the walls

Thought-responsive or biological interface implied

Corso reported briefings suggesting the craft responded to its occupants neurologically

No written language

Symbols reported, but not alphabetic or numeric

Often described as geometric or pictographic

Seating & Crew Area

Small, contoured seats

Moulded directly into the structure

Sized for small-bodied occupants

Minimal crew capacity

Estimated 3–5 beings' maximum

No visible restraints

Seats appeared shaped to support the body naturally

Environmental Systems

No visible life-support machinery

No hoses, tanks, valves, or breathing apparatus

Atmosphere allegedly breathable

No helmets or suits worn inside (per reports Corso cited)

Temperature regulation not apparent

Interior reportedly cool and stable

Glenn Dennis (Mortician / Funeral Home Director, Roswell)

Dennis did not claim to see the interior directly, but reported conversations with base personnel who allegedly did.

Interior Observations (as relayed to him)

Interior smelled "sterile" or "chemical"

Lighting described as diffuse

No visible bulbs or fixtures

Light appeared to emanate from walls or ceiling

No blood or organic debris

Interior described as clean despite crash

Lt. Walter Haut (Affidavit, 2002 – indirect interior references)

Haut did not describe the interior in detail, but relayed what he was told by senior officers.

Craft described as intact internally

Exterior damaged, interior largely preserved

Occupants found inside or near the craft

Interior "unlike any aircraft"

No engines, wiring, or fuel systems recognizable

Nurse-related Accounts (Second-hand, Alleged)

Attributed to unnamed medical or base personnel cited by Dennis and later authors.

Interior Medical/Operational Notes

No storage compartments

No food, water, or waste systems observed

No tools or maintenance equipment

Technology appeared "organic" rather than mechanical

Interior surfaces described as slightly warm or smooth to the touch

Common Descriptive Themes Across Accounts

These details recur across Corso's briefings and witness-derived reports:

Seamless construction

Biological or neurological control systems

Minimalist design (nothing "extra")

Absence of human-style engineering logic

Integration of structure, control, and function

Interior optimized exclusively for its occupants

Composite reconstruction of the interior of the Roswell craft, produced by synthesizing recurring details from eyewitness testimony and second-hand accounts. This illustration reflects consistent descriptions of a compact, seamless interior with curved geometry, integrated controls, moulded seating, and the absence of conventional mechanical systems, as reported by military briefings attributed to Philip J. Corso, indirect observations relayed by Glenn Dennis, statements referenced by Lt. Walter Haut, and other contemporaneous accounts. The image is not presented as photographic evidence, but as a visual aid to convey the common structural and functional features repeatedly described across independent testimonies.

Chapter 24: The Avro Canada Project and the Limits of Human Replication

One of the most intriguing and least understood chapters in post-war aerospace history is the Avro Canada VZ-9AV Avrocar project. Officially, the Avrocar was an experimental vertical take-off and landing (VTOL) aircraft developed during the 1950s and early 1960s, jointly funded by the United States Air Force and the U.S. Army. Its disc-shaped design, radical even by Cold War standards, was intended to explore new possibilities in lift, manoeuvrability, and battlefield deployment. Ultimately, the project was deemed a failure and quietly shelved. Yet it is precisely this failure that warrants closer examination. From a purely conventional perspective, the Avrocar's shortcomings—instability, poor lift, uncontrollable oscillation, and limited altitude—appear to confirm that disc-shaped aircraft were impractical using known propulsion methods. However, this explanation raises a deeper and more troubling question: why was such an inherently unstable configuration pursued so persistently, despite repeated evidence that it could not function as intended?

By the early 1950s, aerospace engineers were well aware of the aerodynamic challenges posed by circular aircraft. Conventional aircraft design had already converged on wings, fuselage, and tail assemblies for reasons rooted in physics and control theory. The Avrocar, by contrast, seemed to reject this accumulated knowledge entirely. Its continued development—despite years of setbacks—suggests that the project was driven by more than theoretical curiosity. This is where the Avrocar begins to intersect with earlier, unresolved events. If reports from Roswell (1947) and Aztec (1948) are even partially accurate—particularly descriptions of disc-shaped craft exhibiting extraordinary manoeuvrability—then the Avrocar takes on a new context. Rather than representing a speculative leap forward, it may represent an attempt to replicate observed performance without fully understanding its underlying mechanisms. In other words, engineers may have been trying to recreate form in the absence of function.

The Avrocar's failure does not necessarily indicate that flying discs were impossible. Instead, it may indicate that human technology at the time was fundamentally inadequate to reproduce what had been observed or recovered. Reverse engineering, by its nature, assumes access to both the artefact and the scientific principles that govern it. Where those principles are unknown—or only partially grasped—the result is often approximation rather than replication.

This perspective reframes the Avrocar not as a misguided experiment, but as a symptom of a larger problem: attempting to force conventional propulsion, materials, and control systems into a configuration that may have relied on entirely different physical principles. If engineers were working from incomplete schematics, recovered fragments, or second-hand descriptions—whether from Roswell, Aztec, or related incidents—the outcome would almost certainly resemble the Avrocar: ambitious, persistent, and ultimately unworkable. Importantly, this interpretation also explains the paradox of the project's longevity. Had the Avrocar been merely an abstract experiment, it would likely have been abandoned far earlier. Its continued funding suggests that decision-makers believed the design should work—that somewhere, evidence existed that it already had.

Failure, in this context, becomes meaningful. It suggests not that the underlying concept was fictional, but that the technological gap between what was observed and what could be reproduced was simply too great. History offers many parallels: when revolutionary technologies are encountered without the scientific framework to support them, imitation often precedes understanding—and fails. Seen through this lens, the Avrocar stands as indirect evidence of something profoundly unsettling. It may represent not the limits of imagination, but the limits of translation—an effort to recreate performance benchmarks that originated outside the known trajectory of human engineering. If Roswell and Aztec introduced materials, configurations, or capabilities that defied contemporary understanding, then the Avrocar's collapse is not an argument against those events. It may, instead, be one of their quietest confirmations.

Footnotes: Avro Canada VZ-9AV Avrocar – Documented Test Results

1. U.S. Air Force Evaluation of Stability and Control

Declassified U.S. Air Force test reports concluded that the Avrocar suffered from severe instability caused by "hubcapping," a rapid oscillation phenomenon that rendered the craft uncontrollable beyond minimal altitude. The vehicle could not maintain stable flight above approximately 3–4 feet without significant pilot correction.

Source:

U.S. Air Force, Project 1794 / Avro VZ-9AV Evaluation Reports, Wright-Patterson AFB, 1959–1961.

2. Maximum Achieved Altitude and Performance Limitations

During repeated test flights conducted between 1959 and 1961, the Avrocar failed to achieve its intended operational ceiling. Official documentation confirms that the craft never exceeded approximately 18 inches to 3 feet of sustained hover in controlled conditions, far below its projected performance parameters.

Source:

U.S. Army Transportation Research Command (TRECOM), Avrocar Flight Test Summary, Fort Eustis, Virginia, 1961.

3. Propulsion Inefficiency and Lift Failure

Engineering assessments identified the turbo-rotor propulsion system as fundamentally inefficient. The airflow required to generate lift created excessive heat, vibration, and instability, while producing insufficient thrust to overcome ground-effect limitations.

Source:

National Research Council of Canada (NRC), Avro Aircraft Ltd. Experimental VTOL Studies, Ottawa, late 1950s.

4. Pilot Testimony and Operational Impracticality

Test pilots reported extreme difficulty controlling the craft, describing it as "dangerously unstable" and "unpredictable," particularly when transitioning out of ground effect. Pilots noted that even minor control inputs often resulted in violent oscillations.

Source:

Paul R. Hill (NASA Langley), pilot evaluation notes cited in USAF Avrocar testing documentation; see also NASA Technical Memoranda, early 1960s.

5. Official Termination of the Program

The Avrocar program was officially terminated in 1961 after joint review by the U.S. Air Force and U.S. Army, citing failure to meet basic flight performance requirements and lack of a viable path forward using existing propulsion and control technologies.

Source:

U.S. Department of Defence, Termination Summary: Avro VZ-9AV, 1961 (declassified).

6. Absence of Aerodynamic Explanation for Disc Stability

Post-program analyses concluded that no conventional aerodynamic solution existed—within then-current scientific understanding—to stabilize a disc-shaped aircraft using reaction thrust alone.

Source:

NASA Historical Office, Experimental VTOL Aircraft of the Cold War Era, archival briefing papers.

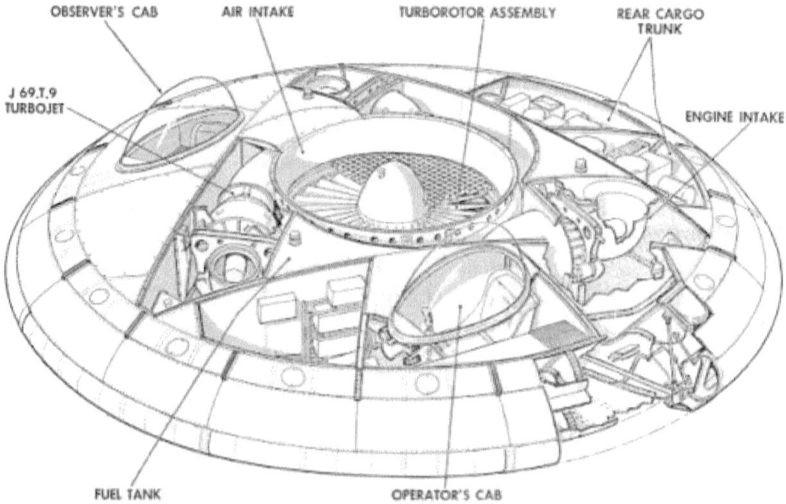
OBSERVER'S CAB
AIR INTAKE
TURBOROTOR ASSEMBLY
REAR CARGO TRUNK
J 69.T.9 TURBOJET
ENGINE INTAKE
FUEL TANK
OPERATOR'S CAB

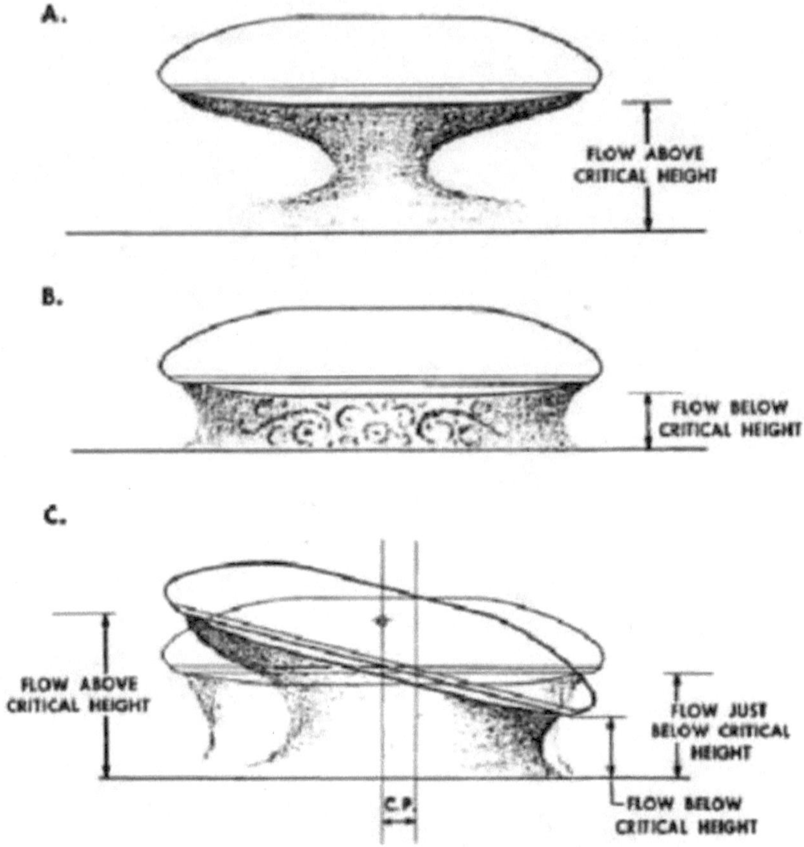
A.
FLOW ABOVE CRITICAL HEIGHT
B.
FLOW BELOW CRITICAL HEIGHT
C.
FLOW ABOVE CRITICAL HEIGHT
FLOW JUST BELOW CRITICAL HEIGHT
FLOW BELOW CRITICAL HEIGHT
C.P.

Image 1: Ground-effect / "critical height" schematic

Caption:

Figure X. Illustration demonstrating ground-effect airflow behaviour and "critical height" limitations in disc-shaped aircraft.

This schematic illustrates airflow instability encountered by circular or disc-shaped vehicles operating close to the ground. Such diagrams were used in mid-20th-century aerodynamic research to explain the loss of lift and control that occurs when a craft rises above a narrow altitude band—an issue that would later prove central to the failure of the Avro Canada VZ-9 Avrocar program.

Reference:

U.S. Army Transportation Research Command (TRECOM). Aerodynamic Studies of Circular Wing Aircraft, c. 1954–1959.

(Also reproduced and discussed in:

Randle, Kevin D. The Roswell UFO Crash Update: 1995. New York: Avon Books, 1995.)

Image 2: Avro Canada VZ-9 "Avrocar" in hover

Caption:

Figure X. Avro Canada VZ-9 Avrocar during hover testing at Wright-Patterson Air Force Base, late 1950s.

Photograph shows the Avrocar prototype hovering a short distance above the ground during test trials. Despite its visually "saucer-like" appearance, the aircraft was unable to exceed approximately three to five feet of stable altitude due to severe control instability and loss of lift outside ground effect.

Reference:

U.S. Air Force Historical Research Agency (AFHRA).

Project VZ-9 Avrocar photographic archives, Wright-Patterson AFB, Ohio.

See also:

Fitzpatrick, Richard. "The Avrocar: A Case Study in VTOL Failure." Air Power History, Vol. 42, No. 3 (1995).

Image 3: Avrocar internal cutaway diagram

Caption:

Figure X. Cutaway technical illustration of the Avro Canada VZ-9 Avrocar showing internal layout and propulsion system.

This diagram details the Avrocar's internal configuration, including the central turbo-rotor assembly, air intakes, fuel tanks, and crew positions. The design highlights the complexity and mechanical density required to approximate vertical lift—contrasting sharply with the reported simplicity and seamless interiors described in UFO crash testimonies from Roswell and Aztec.

Reference:

Avro Aircraft Ltd. of Canada. VZ-9 Avrocar Engineering Drawings and Technical Reports, 1958–1961.

Declassified and held by the National Archives and Records Administration (NARA), College Park, Maryland.

Discussed in:

Randle, Kevin D. The Roswell UFO Crash Update: 1995. Avon Books.

- END -

SHIFT HAPPENS DEPT.

FROM THE DESK OF ______________________, ESQ. DATE: ___/___/20___

Final Chapter: Authors Closing Reflections

Writing From the desk of Christian Stefanizzi

I joined the Mutual UFO Network (MUFON) in November 2025, not to confirm preconceived beliefs, but to seek answers. Like many investigators before me, I began this journey with skepticism, curiosity, and willigness to follow evidence wherever it led. What this research has reinforced is not that governments never tell the truth, but that they often struggle to do so when the truth challenges public stability, institutional authority, or national security narratives.

Those who question why information relating to UFO encounters would be withheld need only look to recent history. Public reactions to shortages whether fuel, food, or basic necessities have shown how quickly fear can override reason. In such moments, trust erodes rapidly, and panic spreads faster than facts. If even minor crises can destabilize public confidence, it is reasonable to consider society might react to confirmation of non human intelligence or advanced techn ology beyond conventional understanding.

For this reason, disclosure if it occurs may never arrive as a single, definitive announcement. Instead, it is more likely to emerge gradually through carefully managed releases of information, terminology shifts, and partial acknowledgements. What was once refered to as UFOs is now officially termed UAP Unidentified Aerial Phenomena. The phenomenon itself has not changed; only the language surrounding it.

Recent whistleblowers and officials, such as those involved in the Pentagons AATIP program, have acknowledged that unidentified objects continue to be investigated. This stands in contrast to the long held claim that Project Blue Book closed the subject in Nineteen sixty nine. The persistence of official interest suggests that the story did not end it merely went quiet.

Roswell remains emblematic of this silence. The explanation has evolved from weather balloon, to secret project, to Project MOGUL each revision mirroring the same structure of denial first established in Nineteen Forty Seven. Yet public awarness has grown. Documents have been declass ified. Witnesses have spoken. Patterns have emerged. While certainty remains elusive, dismissal is no longer sufficient.

This book does not claim to offer final answers. What it does aim to provide is clarity by seperating hoaxes from testimony, speculation from documentation, and assumption from investigation. If nothing else, it demonstrates that the question is not whether something unexplained occured, but why the conversation surrounding it has been so carefully controlled.

History shows that truths buried deeply enough eventually resurface. Whether through archives, testimony, or technological neccesity, they find their way back to the light. The evidence suggests that whatever occured in the New Mexico desert in Nineteen Forty Seven has not been forgotten only deferred.

Christian Stefanizzi
The Roswell Investigator

Addendum (A)

The Author's Final Conclusive Investigation into the Air Force Report: Roswell Case Closed

Christian Stefanizzi, MA

Compiled: 23 December 2025

The Damage and the Legacy of the Alien Autopsy Film

It would be intellectually dishonest to deny that the so called Alien Autopsy footage produced by Spyros Melaris and Ray Santilli has caused measurable damage to the credibility of serious Roswell research. At the same time, paradoxically, it propelled the Roswell incident into global consciousness well into the twenty-first century. Whatever one's position on its authenticity, Roswell has become a modern mythos—one that is unlikely to fade from public memory and will almost certainly persist as a living legend.

In a 1997 interview, respected UFO researcher Kevin D. Randle was asked for his opinion on the film. He reportedly laughed briefly, rolled his eyes, and then stated:

"It's done a lot of damage to legitimate research. But the cameraman—or whoever is claiming to be him—has made Roswell history in another powerful way."

This reaction encapsulates the broader academic dilemma: Roswell is simultaneously damaged by sensationalism and sustained by it.

The Fundamental Question to the United States Government

My central question directed hypothetically to the United States Air Force, the Department of Defence, and the intelligence community is this:

If the Roswell debris was nothing more than a classified Project MOGUL balloon or a Rawin radar target reflector, where is the physical evidence today?

Where is a contemporary photograph or video, in 2025, of the United States government holding the recovered material and stating unequivocally: "This is what all the fuss was about."

Skeptics, including the Air Force itself, counter this by citing The Roswell Report: Case Closed, asserting that Project MOGUL is documented exhaustively and that the matter has therefore been resolved. However, "case closed" is not synonymous with conclusive.

As demonstrated throughout this book, two critical claims within the Air Force report remain unresolved.

The "Alien Bodies" Explanation Examined

The Air Force further attempts to explain eyewitness testimony describing non-human bodies by referencing two unrelated post-Roswell incidents:

A 1956 KC-97G aircraft crash at Walker Air Force Base, resulting in the deaths of eleven airmen.

A 1959 manned balloon mishap associated with Project Excelsior, involving Captain Joseph Kittinger.

Both explanations fail under scrutiny.

The 1956 KC-97 Crash

The KC-97G accident occurred on 26 June 1956 when a propeller blade separated during take-off, struck the fuselage, and caused catastrophic failure. All eleven crew members were killed. According to the Aviation Safety Network, the aircraft was "totally destroyed."

This raises immediate questions:

If the aircraft was totally destroyed, how could identifiable bodies be mistaken for non-human entities?

Why has no photographic or forensic evidence of these remains been presented to support such a comparison?

The Air Force provides no documentation linking this crash to eyewitness descriptions from Roswell. Instead, it appears to function as a post-hoc explanatory placeholder.

The 1959 Project Excelsior Incident

Project Excelsior, conducted on 16 November 1959, involved Captain Joseph Kittinger testing high-altitude parachute systems. During Excelsior I, a malfunction caused Kittinger to lose consciousness while spinning at approximately 120 revolutions per minute. He survived only due to an automatic parachute deployment at 10,000 feet.

Kittinger was recovered by a waiting ground team and returned to Wright-Patterson Air Force Base. His recovery is thoroughly documented and photographed.

However, this incident does not align with Roswell testimony:

Frankie Rowe described four bodies, approximately 3.5–4 feet tall.

She stated the heads were disproportionately large, with almond-shaped eyes, and that the beings wore tight-fitting suits, not flight gear.

Joseph Kittinger was approximately 5'9"–5'10", wore a helmet and pressure equipment, and exhibited no abnormal cranial features or swelling.

There is no anatomical, temporal, or contextual overlap sufficient to support the Air Force's claim.

Why "Case Closed" Remains Open

The Project Excelsior case is exceptionally well documented. Ironically, this abundance of evidence only highlights the absence of equivalent documentation for Roswell. The Air Force explanation does not resolve the eyewitness accounts—it contradicts them.

As of 2025, the Roswell case remains unresolved. Rather than strengthening the Air Force's position, these explanations weaken it. Increasingly, researchers and investigators recognize these explanations as insufficient and evasive.

The pattern that emerges is one of selective disclosure, retroactive justification, and institutional denial.

Final Editorial Assessment

This addendum demonstrates that Roswell Case Closed is not an evidentiary conclusion, but a bureaucratic one. The unresolved absence of records, the reliance on unrelated incidents, and the failure to address consistent eyewitness testimony collectively indicate that the Roswell incident is far more complex than officially acknowledged.

Whether full disclosure will ever occur remains uncertain. What is clear is that the Air Force's explanation does not withstand rigorous scrutiny and history shows that denial often conceals more than it reveals.

Another explanation put forward in the U.S. Air Force's *Roswell Report: Case Closed* is that **anthropomorphic test dummies** were responsible for eyewitness descriptions of "alien bodies." This explanation, upon close examination, is deeply flawed and historically inaccurate. It increasingly appears that the Air Force, rather than resolving inconsistencies, is **clutching at any available explanation that might vaguely resemble the reported descriptions**, regardless of whether it withstands scrutiny.

The fundamental problem with the test-dummy explanation is chronological.

The U.S. Air Force did **not begin using anthropomorphic test dummies in high-altitude balloon experiments until 1953**, fully **six years after the Roswell incident of July 1947**. The earliest documented use of such dummies occurred during projects associated with high-altitude escape and ejection testing in the early 1950s. As such, it is impossible for witnesses in 1947 to have mistaken anthropomorphic test dummies for extra-terrestrial bodies, because **those dummies did not yet exist**.

This alone seriously undermines the Air Force's claim.

The Size Discrepancy

The explanation becomes even more untenable when the **physical dimensions** of the test dummies are compared with eyewitness testimony.

Frankie Dwyer Rowe, daughter of Roswell firefighter Dan Dwyer, stated that her father was dispatched to a crash site outside Roswell on or around **4 July 1947**, where he reportedly observed **four non-human beings**. According to Rowe's testimony, her father described the beings as:

- approximately **3.5 to 4 feet tall**,
- with **small bodies and disproportionately large heads**,
- **grey in colour**,
- possessing **almond-shaped eyes**,
- and wearing what appeared to be **one-piece, skin-tight suits**.

Rowe further stated that one of the beings was **still alive**, making a low moaning sound "as if calling for help."

Now compare this with the first widely documented anthropomorphic test dummy, **Sierra Sam**, developed in **1949**.

According to official sources cited in *Case Closed* itself:

- Sierra Sam was approximately **6 feet tall (183 cm)**,
- weighed around **91 kilograms**,
- was **tan/brown in colour**,
- and wore a **flight helmet and flight suit**.

These characteristics bear **no resemblance whatsoever** to the beings described by Rowe and other eyewitnesses.

Appearance and Function

Anthropomorphic test dummies are, by definition, **human analogues**. They are modelled on human anatomy to simulate the effects of impact, acceleration, and environmental stress on the human body. As such, they do not possess oversized heads, non-human facial features, or unusual proportions. Nor are they designed to resemble anything other than a human form.

There is **no evidence whatsoever** of any test dummy:

- measuring **3.5 to 4 feet in height**,
- coloured **grey**,
- possessing **large almond-shaped eyes**,

- lacking helmets or equipment,
- or having a head size disproportionate to the body.

The Air Force may attempt to counter this by suggesting that earlier, experimental dummies were smaller or "cruder" in design. However, this argument fails for a simple reason: **there is no evidence of any predecessor to Sierra Sam** that matches the descriptions given by Roswell witnesses. No photographs, no documentation, no procurement records, and no engineering notes support the existence of such a device.

A Question of Logic

The core issue here is one of logic.

How could engineers design an anthropomorphic test dummy that resembles something **they had never seen before**—a form that does not correspond to any known human anatomy? Test dummies are created to replicate humans, not hypothetical non-human entities. The beings described by eyewitnesses do not resemble damaged pilots, injured airmen, or experimental mannequins. They resemble **none of the known equipment or personnel used by the U.S. military in 1947**.

The conclusion is unavoidable: **the anthropomorphic test dummy explanation does not fit the evidence.**

It does not fit the timeline.

It does not fit the physical descriptions.

It does not fit the documented history of U.S. Air Force testing programs.

To insist otherwise is not reasoned analysis—it is retroactive justification.

In light of this, the suggestion that the Roswell witnesses mistook test dummies for non-human beings must be rejected. The descriptions given do not correspond to any known human-made object or test apparatus. What was described was **not an anthropomorphic test dummy**, but something entirely outside the known technological and biological framework of the time.

The Air Force's Summary Conclusion and Its Implications

In *The Roswell Report: Fact vs. Fiction in the New Mexico Desert* (1994), the U.S. Air Force presents a summary conclusion regarding the nature of the debris recovered near Roswell, New Mexico, which appears on page 325 of the report. This conclusion has also been highlighted and examined by independent researchers associated with the Ramey Memo investigation, notably through the website *Ramey-Memo.com*. That research includes analysis of the so-called "Ramey Memo" itself, as well as the individuals connected to the well-known Fort Worth Army Air Field photographs, including James Bond Johnson, the photographer who documented the scene showing Major Jesse Marcel and Brigadier General Roger Ramey holding the recovered debris. The memo visible in those photographs has since been examined by researchers such as David Rudiak and others affiliated with *Ramey-Memo.com*, yielding additional insights relevant to the Air Force's conclusions.

Before examining the broader problems with the Air Force's explanation in *Fact vs. Fiction*, it is important to review the report's own stated conclusion regarding the identity of the wreckage. The report states:

"Professor Moore opined that the material was most likely the shredded remains of a multi-neoprene balloon train with multiple radar reflectors. The material and a 'black box,' described by Cavitt, was, in Moore's scientific opinion, most probably from Flight 4, a 'service flight' that included a cylindrical metal son buoy and portions of a weather instrument housed in a box, which was unlike typical weather radiosondes which were made of cardboard. Additionally, a copy of a professional journal maintained at the time by A. P. Crary, provided to the Air Force by his widow, showed that Flight 4 was launched on June 4, 1947, but was not recovered by the NYU group. It is very probable that this Top-Secret project balloon train (Flight 4), made up of unclassified components, came to rest some miles northwest of Roswell, New Mexico, became shredded in the surface winds, and was ultimately found by the rancher, Brazel, ten days later."

The first and most immediate difficulty with this conclusion is that Flight 4 was never recovered and is therefore used retrospectively as a best-fit explanation rather than being identified at the time of the incident. Given Project Mogul's classified role in detecting Soviet nuclear tests, it is reasonable to expect that the loss of such a flight would have generated concern, documentation, or follow-up by the New York University research group, military intelligence, or associated Army Air Forces units. Yet no such contemporaneous evidence has been produced. Moreover, once photographs of the recovered debris appeared in national newspapers—showing Major Marcel and General Ramey at Fort Worth Army Air Field—it would seem likely that NYU personnel or military official's familiar with Mogul hardware would have recognized the material and acknowledged the recovery of the missing flight. The absence of any documented reaction or clarification at that time represents a significant gap in the Air Force's explanation and forms the foundation for the unresolved questions that follow.

This absence of contemporaneous recognition or response leads directly to a second, closely related problem with the Air Force's conclusion: the handling and ultimate disposition of the recovered material itself. If the debris recovered near Roswell was indeed part of a Top-Secret intelligence-gathering project, its treatment after recovery becomes critically important. The official explanation offers no clear account of why the material was allegedly discarded, destroyed, or otherwise allowed to vanish from the historical record, nor does it identify any preserved examples that can be independently examined today. In cases involving classified technology, particularly those tied to strategic nuclear detection, physical evidence would normally be retained for documentation, verification, or future reference.

The Absence of Physical Evidence and the Problem of Verification

A central weakness in the Air Force's Project Mogul explanation lies in the absence of any surviving physical evidence. According to *The Roswell Report: Fact vs. Fiction in the New Mexico Desert*, the debris recovered in 1947 was ultimately discarded, destroyed, or otherwise lost, leaving no material available for independent examination. This is difficult to reconcile with the claim that the wreckage originated from a Top-Secret intelligence program designed to detect Soviet nuclear detonations. In comparable cases involving classified or experimental technology, recovered materials are typically preserved, catalogued, and retained precisely to enable verification, technical analysis, and historical accountability.

The report does not provide documentation explaining who authorized the disposal of the debris, when this occurred, or under what security protocols such an action was taken. Nor does it explain why no representative samples were retained, even after the incident became publicly controversial. The absence of preserved material prevents modern researchers from conducting forensic or materials analysis that could conclusively support or refute the Mogul explanation, thereby placing disproportionate weight on retrospective interpretation rather than verifiable evidence.

Compounding this issue is the reliance on reconstructed timelines and later recollections to identify Flight No. 4 as the most probable source of the debris. While such reconstructions can be useful in the absence of complete records, they cannot substitute for contemporaneous identification or physical proof. The conclusion that Flight 4 drifted northwest of Roswell, fragmented in surface winds, and was recovered by a local rancher remains an inference rather than a documented event. Without preserved debris or contemporaneous acknowledgment linking the recovered material to Project Mogul, the explanation remains circumstantial.

Taken together, the lack of contemporaneous concern over the missing flight, the absence of documentary confirmation following public release of the recovery photographs, and the unexplained disappearance of the physical evidence create a cumulative evidentiary gap. This gap does not in itself establish an alternative explanation for the Roswell incident, but it does undermine the claim that the case has been conclusively resolved. The Air Force's Mogul explanation, while technically plausible, rests on incomplete documentation and assumptions that cannot be independently verified. It is for these reasons that the Roswell incident continues to resist final closure and remains open to continued historical and analytical inquiry.

Individuals Cited as Witnesses to Non-Human Bodies (Roswell, 1947)

A. Claimed Direct or Near-Direct Witnesses (As Reported in Testimony)

These individuals are **frequently cited** as having either seen the bodies themselves or being present at locations where bodies were handled.

1. **Glenn Dennis**
   Mortician, Ballard Funeral Home, Roswell
    - Claimed he was contacted by the base hospital regarding small coffins
    - Stated a nurse friend described non-human bodies at the base hospital
    - His testimony is controversial but central to body claims
2. **Frankie Dwyer Rowe (via her father, Dan Dwyer)**

Dan Dwyer – Roswell Fire Department

- Rowe stated her father was sent to a crash site
- He allegedly described **four small grey beings**, oversized heads, almond-shaped eyes
- This is **second-hand testimony**, but detailed and consistent over decades

3. **Joe Montoya (also spelled Montoya / Mantuia in some sources)**
   Military police, Roswell Army Air Field
   - Reportedly present during body recovery
   - Did not give public interviews himself; cited through secondary sources
4. **Rubén Anaya**
   Associated with base operations
   - Described as deeply shaken
   - Reportedly went home and drank heavily afterward, contemplating what he had seen
   - Again, **secondary testimony**, but consistent with psychological shock patterns

B. Military Personnel Alleged to Have Handled or Guarded Bodies

1. **Pfc. Sheridan Cavitt**
   Counterintelligence Corps (CIC)
   - Officially denied bodies
   - However, some researchers argue his statements changed over time
2. **Arthur Exon**
   Brigadier General, USAF (ret.)
   - Did **not** claim to see bodies himself
   - Stated bodies were recovered elsewhere and flown to

Wright-Patterson
    - Considered a **high-credibility indirect witness**

C. Medical and Technical Personnel (Indirect Witnesses)

1. **Unnamed Army Air Force Nurse (associated with Glenn Dennis)**
    - Allegedly witnessed autopsies
    - Identity never confirmed
    - Existence disputed, but central to body-handling narrative
2. **Lt. Col. Philip J. Corso**
    - Claimed knowledge of bodies via briefings and documents
    - Did not claim first-hand observation
    - Testimony is **indirect but influential**

D. Wright-Patterson Air Force Base–Related Testimony

1. **Dr. Eric A. Walker**
   Scientist, consultant, later President of Penn State
    - Attended meetings (1949–1950) discussing recovered craft and occupants
    - Correspondence suggests **knowledge of non-human entities**, later partially retracted
2. **Dr. Detlev W. Bronk**
   MJ-6 (alleged), biophysicist

- No direct testimony of bodies
- **Notable absence of 1947 documentation** despite meticulous records

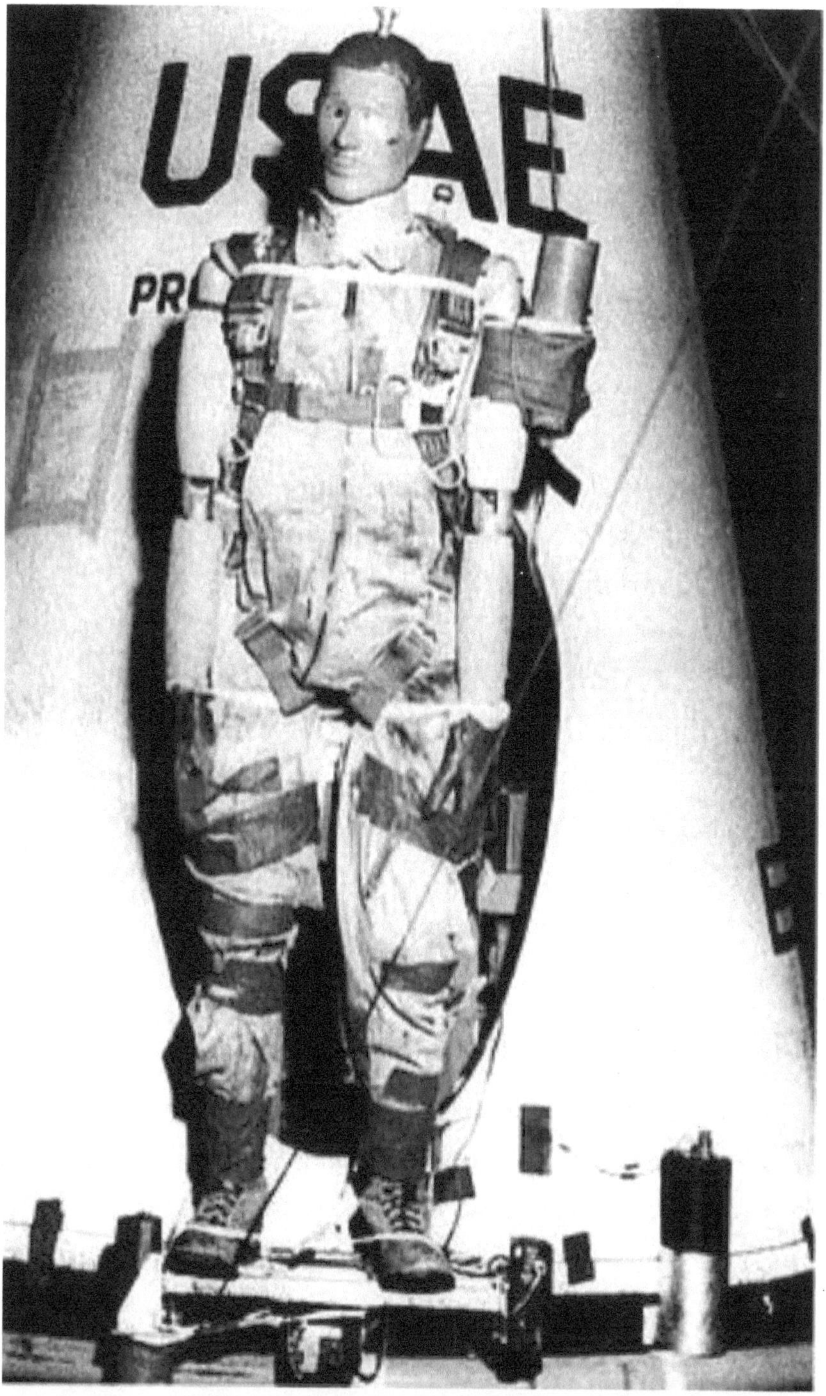

Image 1 (Top): Three Air Force Personnel with Anthropomorphic Test Dummy

Caption (suggested)

U.S. Air Force personnel posing with an Alderson anthropomorphic test dummy used in high-altitude research programs.

Photographs of this nature were later cited by the U.S. Air Force in the 1990s as part of its explanation that eyewitness accounts of bodies associated with the Roswell incident were the result of confusion with post-1950s crash-test dummies.

Reference

U.S. Air Force. The Roswell Report: Case Closed. Washington, D.C.: Headquarters United States Air Force, 1997.

See also:

Holloman Air Force Base Archives, Anthropomorphic Test Dummy Program, 1953–1960.

Image 2 (Bottom): Air Force Test Subject in High-Altitude Suit (High Dive / Excelsior)

U.S. Air Force test subject wearing a high-altitude pressure suit during early human survivability experiments.

Such imagery was later presented by the Air Force as visual context for explaining reports of "non-human bodies," despite these programs post-dating Roswell by several years.

Reference

U.S. Air Force. A History of High-Altitude Escape Systems. Holloman Air Force Base, New Mexico, declassified technical photographs, 1953–1960.

See also:

Project High Dive (1953–1958); Project Excelsior (1959–1960).

Addendum [B] The Twinning Memo September 23, 1947
Christian Stefanizzi MA
Compiled 24th of December 2025

The Twining Memorandum: A Forgotten Smoking Gun

My next phase of investigation into the Roswell case led me to one of the most significant—and curiously overlooked—documents of 1947: **the Twining Memorandum**, authored by General **Nathan F. Twining** on **23 September 1947**.

The memo was declassified and released in **1974 under the Freedom of Information Act (FOIA)**. However, at the time of its release, it was largely ignored by military historians and investigators alike. By the mid-1970s, public and institutional fatigue surrounding Roswell was widespread; document after document had surfaced, only to be dismissed or buried beneath official explanations. As a result, the Twining Memo failed to receive the scrutiny it warranted.

This oversight is difficult to justify.

The Authority of General Nathan F. Twining

General Nathan F. Twining was not a fringe figure. He was one of the most highly decorated officers in United States Air Force history and later became **the first Air Force general to serve as Chairman of the Joint Chiefs of Staff**, holding that position from **1957 to 1960.** Twining was a man whose assessments carried enormous weight within military and intelligence circles.

For such an individual to compile and circulate a memorandum addressing unidentified flying objects is, in itself, extraordinary.

The Contents of the Twining Memo

Dated **23 September 1947**, the memo opens with a statement that is both direct and unambiguous:

"Flying disks are real and not imaginary or fictitious."

This alone represents a stunning admission—particularly given that it appears less than three months after the Roswell incident.

The memo proceeds to outline several key conclusions:

1. It is the opinion that:

A. The phenomenon reported is something real and not visionary or fictitious.

B. There are objects, probably approximating the shape of a disc, of such appreciable size as to appear as large as man-made aircraft.

C. Some incidents may be caused by natural phenomena such as meteors.

D. The reported operating characteristics—extreme rates of climb, manoeuvrability (particularly in roll), and evasive action when sighted or contacted by friendly aircraft or radar—suggest that some of these objects are controlled either manually, automatically, or remotely.

E. The apparent common description of the object is as follows:

1. A metallic or light-reflecting surface.

Notably, the memo is stamped **TOP SECRET** at both the top and bottom of the document.

Why This Memo Matters

The Twining Memo is significant for several reasons:

1. **It confirms internal military concern**
   The memo clearly establishes that senior Air Force leadership regarded flying discs as real, physical objects requiring serious investigation.
2. **It describes performance beyond known 1947 technology**
   The extreme manoeuvrability, rapid climb rates, and evasive behaviour described were not consistent with any aircraft publicly or secretly acknowledged in 1947.

3. **It explicitly uses the term "disc"**
   This is critical. The use of "disc" aligns directly with descriptions found in the **Ramey Memo**, dated July 1947, which—based on legible portions—appears to reference a recovered "disc" and "victims" being transported to Fort Worth, Texas.

The linguistic consistency between the Twining Memo (September 1947) and the Ramey Memo (July 1947) is striking. Two separate documents, authored within months of one another, both refer to disc-shaped craft—undermining later claims that the events involved weather balloons, Project MOGUL, or RAWIN radar reflectors.

A Reasoned Conclusion

Based on the Twining Memo alone, it becomes increasingly difficult to maintain the official explanation that the Roswell incident involved nothing more than conventional or experimental balloon technology. The memo describes objects exhibiting characteristics far beyond the capabilities of any known aircraft at the time.

My conclusion, drawn from this document, is cautious but firm:

A disc-shaped craft was in the possession of the United States Air Force in 1947—or at the very least, senior leadership believed such craft existed.

Whether that craft remains in government custody today is unknown. However, the Twining Memo strongly suggests that the phenomenon was taken seriously at the highest levels of military command and that the public explanations offered in later decades are, at best, incomplete.

In the next section—**Addendum C**—I will examine the **Ramey Memo** in detail and explore its implications alongside the Twining Memorandum.

U.S. Army Air Forces Intelligence memorandum titled "AMC Opinion Concerning 'Flying Discs,'" dated 23 September 1947.

The document reflects early post-Roswell internal assessments concluding that reported flying disc phenomena were "real and not visionary or fictitious," and notes performance characteristics exceeding known aircraft capabilities. Originally classified SECRET.

Reference (Chicago / academic style)

U.S. Army Air Forces, Air Materiel Command. AMC Opinion Concerning "Flying Discs". Intelligence memorandum, TSIDIN/DM-1 & G-4-100, Wright Field, Ohio, 23 September 1947. Declassified.

National Archives and Records Administration (NARA), Record Group 341, Entry UD-341, Project Sign / AMC Intelligence files.

SECRET
C O P Y

TSDIN/HMM/ig/6-4100

23 September 1947

TSDIN

SUBJECT: AMC Opinion Concerning "Flying Discs"

TO: Commanding General
Army Air Forces
Washington 25, D. C.
ATTENTION: Brig. General George Schulgen
AC/AS-2

1. As requested by AC/AS-2 there is presented below the considered opinion of this Command concerning the so-called "Flying Discs". This opinion is based on interrogation report data furnished by AC/AS-2 and preliminary studies by personnel of T-2 and Aircraft Laboratory, Engineering Division T-3. This opinion was arrived at in a conference between personnel from the Air Institute of Technology, Intelligence T-2, Office, Chief of Engineering Division, and the Aircraft, Power Plant and Propeller Laboratories of Engineering Division T-3.

2. It is the opinion that:

a. The phenomenon reported is something real and not visionary or fictitious.

b. There are objects probably approximating the shape of a disc, of such appreciable size as to appear to be as large as man-made aircraft.

c. There is a possibility that some of the incidents may be caused by natural phenomena, such as meteors.

d. The reported operating characteristics such as extreme rates of climb, maneuverability (particularly in roll), and action which must be considered evasive when sighted or contacted by friendly aircraft and radar, lend belief to the possibility that some of the objects are controlled either manually, automatically or remotely.

e. The apparent common description of the objects is as follows:

(1) Metallic or light reflecting surface.

Incl. #1

SECRET
COPY

U-39552

Addendum [C], The Ramey Memo Dated the 8th of July 1947.
By Christian Stefanizzi, MA
Compiled on the 27th of December 2025

The most important and incriminating piece of evidence for the entire Roswell case is the Ramey memo. The Ramey memo may be the ultimate key to unlocking the entire case as many scholars and Roswell researchers believe what is on the memo explains whether this was just a MOGUL balloon or a flying disk as the original report released by lieutenant Walter Haut described it. On July 8th 1947, Lieutenant Walter G Haut released a statement to the press stating these exact words as ordered by Colonel William Blanchard base commander;

Headline July 8th, 1947

'The Army Air forces has announced that a flying disc has been found and is now in the possession of the Army. Army officers say the missile found sometime last week has been inspected at Roswell New Mexico and sent to Wright Field Ohio for further inspection'.

That is the original press release from July 8th 1947, as issued by Walter Haut from a Roswell radio news clip which was purchased by myself on Old Time Radio catalogue as it's difficult to get an original recording.

Now we have mentioned the Press release let's move forward a day to July 9th, 1947 and the intersection of the Ramey Memo as the press release will help link the two. The events begin like this, General Roger.M.Ramey is sent to Roswell to relinquish command from Colonel William Blanchard in which the flying disc statement is retracted and a new news headline is released this is seen below in multiple headlines;

'July 9, 1947 – Fort Worth / Nationwide Headlines

"General Ramey Empties Roswell Saucer"

— Fort Worth Star-Telegram, July 9, 1947

"Ramey Says Flying Disc Was Weather Balloon"

— Associated Press, July 9, 1947

"Roswell 'Flying Disk' Is Explained as Weather Balloon"

— Dallas Morning News, July 9, 1947

"Army Debunks Roswell Flying Disc Report"

— Various regional papers via AP and UP wire services

"Recovered Object Not a Flying Disc, Army Says"

— Chicago Tribune, July 9, 1947

Now please find attached below the Ramey memo analysed in its most complete form from what is legible, this has been analysed by four different AI programs:

Enhanced Transcription and Analysis of the Ramey Memo (8 July 1947)

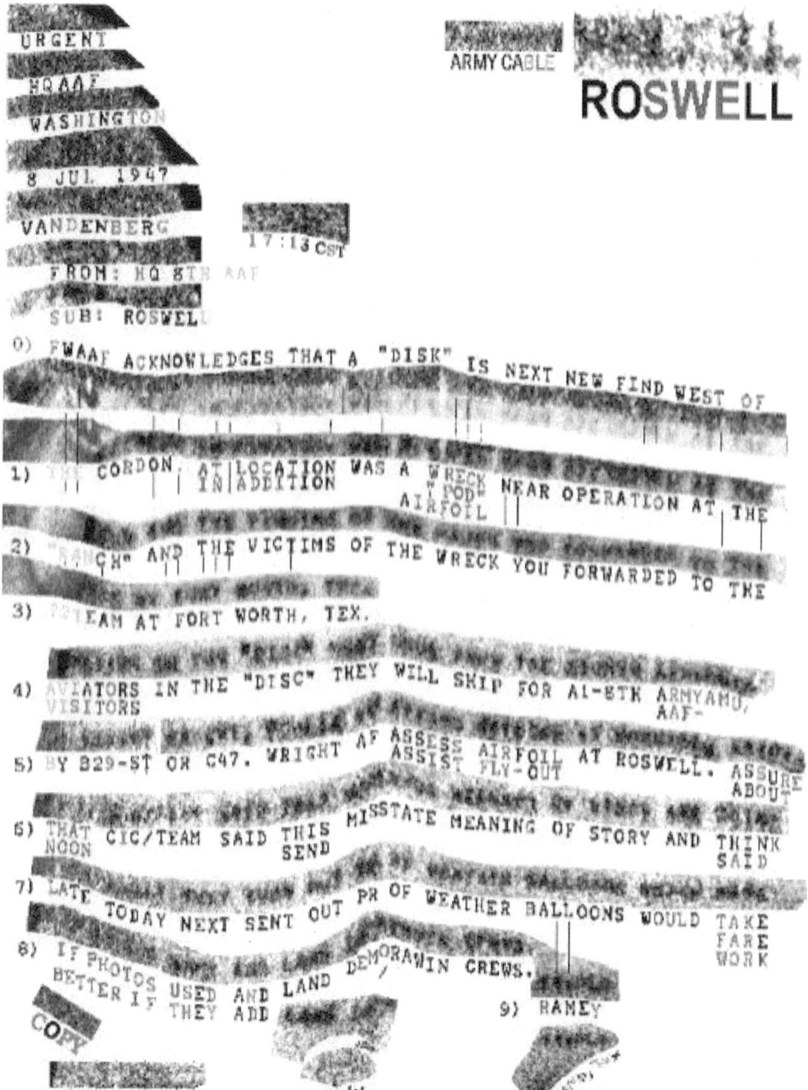

URGENT
HQAAF
WASHINGTON
8 JUL 1947
VANDENBERG
17:13 CST
FROM: HQ 8TH AAF
SUB: ROSWELL

ARMY CABLE
ROSWELL

0) FWAAF ACKNOWLEDGES THAT A "DISK" IS NEXT NEW FIND WEST OF
1) THE CORDON AT LOCATION WAS A WRECK IN ADDITION "POD" NEAR OPERATION AT AIRFOIL THE
2) "RANCH" AND THE VICTIMS OF THE WRECK YOU FORWARDED TO THE
3) TEAM AT FORT WORTH, TEX.
4) AVIATORS IN THE "DISC" THEY WILL SHIP FOR A1-8TH ARMYAMU, VISITORS AAF-
5) BY B29-ST OR C47. WRIGHT AF ASSESS ASSIST AIRFOIL FLY-OUT AT ROSWELL. ASSURE ABOUT
6) THAT CIC/TEAM SAID THIS NOON SEND MISSTATE MEANING OF STORY AND THINK SAID
7) LATE TODAY NEXT SENT OUT PR OF WEATHER BALLOONS WOULD TAKE FARE WORK
8) IF PHOTOS USED AND LAND BETTER IF THEY ADD DEMORAWIN CREWS.
9) RAMEY

COPY
TOP SECRET

Copyright 2009, David Rudiak

Source: Photograph taken in the office of Brig. Gen. Roger M. Ramey, Fort Worth Army Airfield, 8 July 1947, during the public presentation of the Roswell debris. Copyright David Rudiak 2009.

Subject of analysis: The document ("Ramey Memo") held in the hand of Col. Thomas J. Dubose.

This appendix contains the best-possible textual reconstruction based on high-resolution digital enhancement and line-by-line optical analysis.

A1. Confidence Scale

To clarify the reliability of each transcription segment:

High confidence – letters or words clearly readable

Medium confidence – mostly readable but partially obscured

Low confidence – very unclear; reconstruction based on fragmentary marks only

A2. Header Transcription

Line Text Confidence

1 URGENT High

2 HQAAF High

3 WASHINGTON High

4 8 JUL 1947 High

5 VANDENBERG High

6 FROM: HQ 8TH AAF High

7 SUB: ROSWELL High

A3. Body Text – Line-by-Line Reconstruction

(1)

"FWAAF ACKNOWLEDGES THAT A 'DISK' IS ... FIND WEST OF ..."

Confidence: Medium

(Several words missing due to smearing)

(2)

"THE CORDON LOCATION WAS A WRECK ... NEAR OPERATION AT THE ..."

Confidence: Medium

Faint fragments beneath:

"POD" — Low

"AIRFOIL" — Low

(3)

"TEAM AT FORT WORTH, TEX."

Confidence: High

(4)

"AVIATORS IN THE 'DISC' THEY WILL SHIP FOR A1–8TH ARMY AIR ..."

Confidence: Medium / Low

(Ending heavily smudged)

(5)

"BY B29-ST OR C47. WRIGHT AF ASSESS AIRFOIL AT ROSWELL. ASSURE ..."

Confidence: Medium

(6)

"THAT CIC/TEAM SAID THIS MISSTATE MEANING OF STORY AND THINK ..."

Confidence: Medium / Low

(7)

"LATE TODAY NEXT SENT OUT PR OF WEATHER BALLOONS WOULD TAKE ..."

Confidence: Medium

(8)

"IF PHOTOS USED AND LAND ... CREWS."

Confidence: Low

(Only partial fragments discernible)

(9)

"RAMEY"

Confidence: High

Bottom Marking

"TOP SECRET"

Confidence: High

A4. Clearest Fully Readable Line

From the enhanced crop:

> "AND THE VICTIMS OF THE WRECK IN THE 'DISC' THEY WILL SHIP"

Confidence: High

This is the most legible and consequential phrase in the document.

QR code linking to a publicly available interview with Frankie Rowe, who recounts her recollections of her father's involvement with debris recovered near Roswell in 1947.

Included for reference and historical context. The author is not affiliated with YouTube or the video's uploader, and inclusion does not imply endorsement.

QR code linking to a publicly available interview with Walter G. Haut, Public Information Officer at Roswell Army Air Field in 1947, discussing his role in issuing the original "flying disc" press release and his later reflections on the incident.

QR code links to a publicly available interview with Colonel Philip J. Corso, hosted on YouTube. Included for reference and research purposes only.

QR code linking to a publicly available 1978 interview with Major Jesse A. Marcel, former intelligence officer of the 509th Bomb Group, discussing his recollections of the 1947 Roswell incident. Included for reference and historical context.

COPIES DESTROYED
FILE CLOSED

Authority:

Date: 18 NOV 1964

Master Bibliography
Beyond Roswell: Cold War Cover-Ups and Conspiracy Theories

Primary Government & Military Documents

Central Intelligence Agency (CIA). (1953). Report of the Scientific Advisory Panel on Unidentified Flying Objects (The Robertson Panel). Declassified.

Central Intelligence Agency (CIA). (1997). CIA's Role in the Study of UFOs, 1947–1990. Center for the Study of Intelligence. Declassified.

Federal Bureau of Investigation (FBI). (1947–1954). UFO and Flying Disc Memoranda. FBI Vault (Declassified).

General Accounting Office (GAO). (1995). Report to the Honourable Steven H. Schiff: Records Relating to the Roswell Incident. Washington, D.C.

U.S. Air Force. (1947). Press Release: Recovery of a Flying Disc. Roswell Army Air Field.

U.S. Air Force. (1994). The Roswell Report: Fact vs. Fiction in the New Mexico Desert. Headquarters USAF.

U.S. Air Force. (1997). The Roswell Report: Case Closed. Headquarters USAF.

U.S. Department of Defence. (1947–1952). Project Mogul Technical and Flight Records. Declassified selections.

Contemporary Newspaper & Media Sources (1947–1950)

Associated Press. (1947, July 8–9). Army Recovers Flying Disc in New Mexico.

Fort Worth Star-Telegram. (1947, July 8). Gen. Ramey Displays Weather Balloon.

Roswell Daily Record. (1947, July 8–9). RAAF Captures Flying Saucer; Rancher Finds Wreckage.

Farmington Daily Times. (1950, March 18). Flying Saucers Jam New Mexico Skies.

Life Magazine. (1950, June). Photographs of Flying Saucer Taken by Oregon Farmer.

The Oregonian. (1950, June). Strange Object Photographed Near McMinnville.

Scholarly & Academic Works (Cold War, Secrecy, Intelligence)

Gaddis, J. L. (2005). The Cold War: A New History. Penguin Press.

Launius, R. D. (2010). Roswell and the Public Imagination. Smithsonian National Air and Space Museum.

Masco, J. (2006). The Nuclear Borderlands: The Manhattan Project in Post-Cold War New Mexico. Princeton University Press.

McMahon, R. J. (2013). The Cold War: A Very Short Introduction. Oxford University Press.

Swords, M. D. (2017). UFOs, the Military, and the Early Cold War. Journal of Scientific Exploration.

Young, J. M. (1998). Project Mogul: A Critical Historical Assessment. Air Force Historical Research Agency Papers.

Investigative & Analytical Roswell Literature

Berlitz, C., & Moore, W. L. (1980). The Roswell Incident. Grosset & Dunlap.

Carey, T., & Schmitt, D. (2007). Witness to Roswell. New Page Books.

Dolan, R. M. (2000). UFOs and the National Security State, Vol. 1. Hampton Roads.

Friedman, S. (1997). Top Secret / Majic. Marlowe & Company.

Pflock, K. T. (2001). Roswell: Inconvenient Facts and the Will to Believe. Prometheus Books.

Randle, K. D. (1995). The Roswell UFO Crash Update. Fund for UFO Research.

Randle, K. D., & Schmitt, D. (1991). UFO Crash at Roswell. Avon Books.

Eyewitness Testimony & Oral History Sources

Haut, W. G. (2002). Affidavit on the Roswell Incident. Public release.

Marcel, J. A. Jr. (1978–1994). Recorded interviews and testimony.

(Referenced via Randle & Schmitt; Pratt interviews; WWL-TV archive.)

Rowe, F. D. (1990s). Roswell Firefighter Testimony. Oral history interviews.

Rickett, L. (1991). Testimony recorded in UFO Crash at Roswell.

Shirkey, R. (1980). Interview in The Roswell Incident.

McMinnville Photographs & Photographic Analysis

Haines, R. F. (1982). Photo Analysis of the McMinnville UFO. Center for UFO Studies.

Klass, P. J. (1974). UFOs Explained. Random House.

(Skeptical analysis of McMinnville images.)

Maccabee, B. (1976). Optical Analysis of the McMinnville Photographs. Ground Saucer Watch.

Reverse Engineering & Technology Claims

Corso, P. J. (1997). The Day After Roswell. Pocket Books.

Kapany, N. (1960). Fiber Optics: Principles and Applications. Academic Press.

Naval Ordnance Laboratory. (1959). Nickel-Titanium Alloy Research (Nitinol). Declassified technical reports.

Soviet & Comparative Cold War Sources

CIA. (1970s–1980s). Soviet UFO Reports and Intelligence Assessments. Declassified summaries.

Shklovsky, I. S., & Sagan, C. (1966). Intelligent Life in the Universe. Holden-Day.

Correspondence & Original Research

Pope, N. (2025). Personal correspondence with author, November 2025.

Randle, K. D. (2025). Personal correspondence with author, November 2025.

Stefanizzi, C. (2025). Author interviews, archival synthesis, and analytical reconstructions.

[1] airandspace.si.edu. (n.d.). Roswell Daily Record | National Air and Space Museum. Roswell Daily Record. "RAAF Captures Flying Saucer on Ranch in Roswell Region." July 9, 1947, front page. Roswell, New Mexico. *Roswell Daily Record*, July 8, 1947, Vol. 53, No. 156. Roswell, New Mexico.

[2] A photograph of the Alderson type anthropomorphic dummies, which the Air Force claims to be the cause of the eyewitness testimonies description of alien bodies. The Roswell Report: Case Closed (1997), P.g, 34.

[3] Berlitz, Charles, and William L. Moore. The Roswell Incident. New York: Grosset & Dunlap, 1980.Blazey, Laurence. "Drawing of alleged alien survivor escorted by military policemen." In The Roswell Incident, by Charles Berlitz and William L. Moore, p. __. New York: Grosset & Dunlap, 1980. "Drawing by Laurence Blazey, UFO Information Network (UFOIN), P.O. Box 5012, Rome, OH 44085.")

[4] A recreation of the account of Dr. Jesse Marcel Jr. who described handling a small metallic I-beam recovered from the debris, noting its remarkable weightlessness and the presence of "purplish, geometric symbols" along its surface. His account appears in The Roswell Legacy (Forge Books, 2007), pp. 49–53. The reference photographs comes from the UFO museum in Roswell.

[5] A photograph of , LT Eileen.M.Fanton, from the Roswell Report: Case Closed (1997), pg.82.

[6] A photograph of Matilda' O'Donnell Macelroy, from the book titled Alien Interview by Lawrence Spencer (2008).

[7] United States Air Force. (1997). The Roswell Report: Case Closed. Washington, D.C.: Department of the Air Force. (See discussion of Major Dan D. Fulgham and illustrative photographs used to address witness misidentification.)

1. [8] U.S. Air Force, The Roswell Report—Case Closed (1997).

[9] Howe, Linda Moulton. An Alien Harvest: Further Evidence Linking Animal Mutilations and Human Abductions to Alien Life Forms. LMH Productions, 1989.

[10] Bishop, Greg. Project Beta: The Story of Paul Bennewitz, National Security, and the Creation of a Modern UFO Myth. Paraview Pocket Books, 2005.

[11] Moore, William L. "UFOs and the U.S. Government: The Insider's View."

Presentation at the MUFON International UFO Symposium, Las Vegas, NV, July 1, 1989.

(Transcript published in: MUFON UFO Journal, issues from late 1989.)

[12] Dwyer, Kevin. "Disinformation and the Construction of UFO Mythology: Reassessing the Bennewitz Affair." Journal of UFO Studies, vol. 7, 2006, pp. 41–62.

[13] Schmitt, Donald R., and Thomas Carey. Witness to Roswell. New Page Books, 2007.

[14] Phillips, Ted. "Affidavit on Humanoid Hand Sketch." In Leonard H. Stringfield, UFO Crash/Retrievals: Status Report III, MUFON & CUFOS, 1978.

[15] Stringfield, Leonard H. Situation Red: The UFO Siege. New York: Bantam, 1977.

[16] Moore, William L., and Charles Berlitz. "Alien Humanoid Photograph Note." The Roswell Incident. New York: Grosset & Dunlap, 1980, pp. 128–129.

[17] Randle, Kevin D., and Donald R. Schmitt. UFO Crash at Roswell. New York: Avon Books, 1991.

[18] Berlitz, Charles, and William L. Moore. The Roswell Incident. New York: Grosset & Dunlap, 1980.

[19] United States Air Force. The Roswell Report: Case Closed. Washington, D.C.: Government Printing Office, 1997.

[20] Berlitz, Charles, and William L. Moore. The Roswell Incident. New York: Grosset & Dunlap, 1980.

[21] Meade Layne memo (quoted in Berlitz & Moore), pp. 97–100.

[22] Testimony of Dr. Weisberg (ibid., pp. 97–98).

[23] Transport route account (ibid., pp. 98–99).

[24] Baron Nicholas von Poppen testimony (ibid., p. 99; originally quoted in Gray Barker, UFO Report, May 1977).

[25] "William S. to Thomas Townsend Brown, September 17, 1982." Unpublished letter. Reproduced online via UFO archival forums (accessed 2025).

[26] Spencer, Lawrence R. Alien Interview: Based on the Personal Notes and Interview Transcripts Provided by the Late Matilda O'Donnell MacElroy. Alien Interview Publishing, 2008.

[27] Light, Gerald. Letter to Meade Layne, dated 16 April 1954. Reproduced in:

Berlitz, Charles, and William L. Moore. The Roswell Incident. New York: Grosset & Dunlap, 1980, pp. 122–125.

[28] Smith, Wilbert B. Memorandum to the Controller of Telecommunications, Department of Transport, Ottawa, 21 November 1950.

Declassified and downgraded from "Top Secret" to "Confidential" on 15 September 1969.

Published in: Berlitz, Charles, and William L. Moore. The Roswell Incident. New York: Grosset & Dunlap, 1980, pp. 118–119.

[29] Federal Bureau of Investigation. (1947, July 8). Teletype Communication: FBI Dallas to Director and SAC Cincinnati, Subject – Flying Disc, Information Concerning. (Reproduced in Berlitz & Moore, 1980, pp. 142–143).

[30] Berlitz, C., & Moore, W. L. (1980). The Roswell Incident (pp. 140–144). New York: Grosset & Dunlap.

— Includes the 8 July 1947 FBI Dallas Teletype, the 15 July 1947 Hoover memorandum, and Bureau Bulletin No. 42 dated 30 July 1947.

[31] Federal Bureau of Investigation. (1947, July 30). Bureau Bulletin No. 42, Series 1947: Flying Discs. (Reproduced in Berlitz & Moore, 1980, p. 144).

[32] Hoover, J. E. (1947, July 15). Memorandum for Mr. Ladd re: Flying Discs and Army Air Forces Cooperation. (Reproduced in Berlitz & Moore, 1980, p. 140).

[33] Randle, K. D. (1995). Roswell UFO Crash Update. Galde Press.

— See pp. 14–16 on eyewitness sightings by the Franciscan nuns, William Woody, and Jim Ragsdale.

— See pp. 55–56 on problems with the Mogul explanation and Col. Thomas DuBose's confirmation of the cover story.

[34] "A privately shared article supplied directly to the author by Philip Mantle (2025).

[35] A privately shared article supplied directly to the author by Philip Mantle (2025), Roswell solved? By Phillip Mantle.

[36] Roswell solved? By Phillip Mantle.

[37] Randle, Kevin D. Roswell UFO Crash Update. New Brunswick, NJ: Inner light Publications, 1995.

[38] KPIX-TV, Interview with Major Jesse Marcel, conducted by Stanton T. Friedman (1979, aired 1980).

[39] Friedman, Stanton & Berliner, Don. Crash at Corona: The Definitive Study of the Roswell Incident. Paraview Press, 1992.

[40] Ibid.

[41] Ibid.

[42] Marcel, Jesse Jr. Affidavit submitted to the Fund for UFO Research, 1991.

[43] Marcel Jr., Jesse. The Roswell Legacy. New Page Books, 2010, pp. 52–53.

[44] Clarkson, J. (2025). Meeting Roswell Witness Frankie Rowe – James Clarkson – UFO Detective. [online] Ufodetective.com. Available at: https://ufodetective.com/ufo-news/meeting-roswell-witness-frankie-rowe/ [Accessed 1 Dec. 2025].

[45] KPIX-TV, Interview with Major Jesse Marcel, conducted by Stanton T. Friedman (1979, aired 1980).

[46] Marcel Jr., Jesse. The Roswell Legacy. New Page Books, 2010, pp. 52–53

[47] Randle, K.D. (1995). Roswell UFO Crash Update. Inner Light - Global communications.

[48] Clarkson, J. (2025). Meeting Roswell Witness Frankie Rowe – James Clarkson – UFO Detective. [online] Ufodetective.com. Available at: https://ufodetective.com/ufo-news/meeting-roswell-witness-frankie-rowe/ [Accessed 1 Dec. 2025].

[49] Ruffener, K. (n.d.). CIC Records: A Valuable Tool for Researchers.

[50] Corso, P.J. and Birnes, W.J. (2017). The day after Roswell. New York: Gallery Books

[51] Corso, P.J. and Birnes, W.J. (2017). The day after Roswell. New York: Gallery Books

[52] McAndrew, James. The Roswell Report: Case Closed: Enriched edition. Unveiling Truths: Unravelling the Roswell UFO Mystery. Good Press, 2021.

[53] Hesemann, Michael, and Philip Mantle. 1998. Beyond Roswell: The Alien Autopsy Film, Area 51, & the U.S. Government Coverup of UFOs. London: Michael O'Mara Books.

[54] John W. R. Taylor, Rockets and Missiles (London: Hamlyn, 1972), page 135.

[55] Earth files Podcast with Linda Moulton Howe, Ep 160: Is this layered Bismuth-Magnesium-Zinc a metal "skin" from a UFO?

[56] Reiter, Nicholas A., and Lori Schillig. The Bridges of Avalon: Science, Spirit, and the Quest for Unity. iUniverse, 2005.

[57] "About the Roswell Photo Collection | Roswell UFO Incident TEST." n.d. Libraries.uta.edu. https://libraries.uta.edu/roswell/images.

[58] "UFOs and the Guy Hottel Memo." n.d. Federal Bureau of Investigation. https://www.fbi.gov/news/stories/ufos-and-the-guy-hottel-memo.

[59]

[60] Modern illustrative reconstruction of a purported Roswell-type craft. This image is speculative and provided for comparative visualization only; it does not originate from U.S. military archives.

[61] U.S. Air Force
The Roswell Report: Fact vs. Fiction in the New Mexico Desert
Headquarters United States Air Force, 1994
(also reused in **The Roswell Report—Case Closed**, 1997)

[62] Smithsonian National Air and Space Museum (NASM). *Apollo Command Module (various missions), museum exhibition photography*. Washington, D.C.
NASA Image and Video Library, Apollo Program Collection.

[63] NASA / Smithsonian National Air and Space Museum.

Ablative heat shield surface, Apollo Command Module.
Catalogue ID: NASM-A19731423008 (Smithsonian collections).
[64] Randle, Kevin D., Donald R. Schmitt, and additional collaborators.
[65]

www.ingramcontent.com/pod-product-compliance
Lightning Source LLC
LaVergne TN
LVHW100258210926
848480LV00003B/953
* 9 7 9 8 2 3 3 1 8 8 0 4 6 *